Windpower Workshop
Building Your Own Wind Turbine

Hugh Piggott

Centre for
Alternative
Technology
Publications

© Hugh Piggott, 1997, 2000

The Centre for Alternative Technology Machynlleth, Powys, SY20 9AZ, UK
•**Tel**. 01654 705950 •**Fax**. 01654 702782
•**email:** info@cat.org.uk •**Website:** www.cat.org.uk

Published by CAT Publications, CAT Charity Ltd. Registered charity no. 265239.

Assistance Bob Todd, Jules Lesniewski, Tim Kirby, KB Jones,
RG Wilson, Tony Lockwood, Rick Dance, Mark Epton, Dave Thorpe
Illustrations Graham Preston and Hugh Piggott
Cover Photograph Hugh Piggott

ISBN 10: 1-898049-27-0
ISBN 13: 978-1-898049-27-2

5 6 7 8 9

Mail Order copies from: Buy Green By Mail   Tel. 01654 705959

The details are provided in good faith and believed to be correct at the time of
writing, however no responsibility is taken for any errors. Our publications are
updated regularly; please let us know of any amendments or additions which you
think may be useful for future editions.

Printed in Great Britain on 100% recycled Cyclus Print paper
by **Antony Rowe** 01249 659705

Originally published with the invaluable assistance of Environment Wales.

The Author

Hugh Piggott runs his own successful windpower business from his home on the beautiful, appropriately windswept, peninsula of Scoraig, off the coast of Scotland. There he advises individuals and companies at home and abroad on small to medium scale windpower turbines and systems. He has been making windmills for twenty years from scrap parts and teaching others how to do so, for example on the Centre for Alternative Technology's twice-yearly windpower course. His books are amongst the Centre's best sellers. He has a wife and two children. He has also written *It's A Breeze! A Guide to Choosing Windpower, Scrapyard Windpower* and *Choosing Windpower* for C.A.T. Publications.

Contents

Foreword

Windpower bites. It's one of those multi-faceted subjects which appeals to all sorts of people from all sorts of angles - free power, no environmental costs, not to mention a world of opportunity for the 'gadget' folk. But what's it all about, and what can practically be done?

Although wind is one of civilisation's oldest forms of mechanical power, it suffered something of a relapse from the start of this century as the benefits of mass cheap energy supply came through. But as the true (and horrendous) costs of mass fossil fuel use come to light, wind is making a big comeback, particularly with large grid connected turbines in Europe. A growing number of countries are doing their best to encourage wind energy generation as part of a range of vital measures towards sustainability.

Although it may seem contrary to 'green' thinking, in many ways the 'grid' is a useful environmental tool – it allows one area's surpluses to meet another area's needs. This avoids much of the costly (economically and environmentally speaking) requirement for energy storage – common in most small 'off the grid' systems. If we all wanted to use autonomous wind energy with battery storage, we would run out of lead before we got very far!

But there certainly are places where the grid does not reach, and plenty of people wanting to live in them. Scoraig, Hugh Piggott's base in Scotland, is just such a place. With a healthy community heading for a hundred souls (an achievement after total abandonment by the old crofting community in the 1950s), all off the grid, and many with their own small windmills, it has been the perfect place for earning the experience and reputation as the best there is in 'DIY' windpower.

Wind can be an excellent choice for isolated power supply, and such is the nature of the folk who live in such places that many will prefer to 'DIY' wherever possible. This book does an admirable job in filling a gaping hole in the available literature on practical small scale wind engineering. It comes from a disciplined and highly trained devotee who has explored all the angles and learned most of the lessons (many of them the hard way). Wind is no more simple than it is free (again economically and environmentally speaking), and a guide is most recommended. There are plenty of pitfalls, most of them easily avoided.

The reward is that wind is the closest thing to being able to 'magic' clean energy from thin air, and Hugh Piggott is a true guru of the art. Read on and enjoy.

Tim Kirby
Chairman
British Wind Energy Association

Chapter One
A Wild Resource

This book is written for those who want to build their own windmill, and also for those who love to dream. It was inspired by the windpower course at the Centre for Alternative Technology, an event where folk from all backgrounds come together to share the excitement of learning about windpower. Much has been omitted for lack of space, but you can find it elsewhere. For basic knowledge of electricity, forces, and turning moments, look through a school physics book. For details of how to site a windmill and live with windpower, see the companion volumes called *It's a Breeze* and *Off the Grid* (also available from CAT Publications).

The wind: a wild resource

Wind energy is wild stuff, and very tricky to handle. Capturing wind energy is like riding an antelope, when we could be using a Volkswagen. Most newcomers to wind energy underestimate the difficulties. Do not expect to get much useful power from a small windmill in a suburban garden, nor to knock together a reliable windpower system in an afternoon!

Look hard at the size of wind machine needed to produce the energy needed. Is this a realistic project for you? Do you have the workshop facilities? Do you have access to a suitable site, where there is space to allow the machine to operate safely and unobstructed?

If your motivation is to clean up the environment, then small scale windpower is not necessarily the best approach. Insulating your house may well save more energy. If you are an urban supporter of renewable energy, you can ask your local electricity board about their green tarriff, or devote yourself to winning the environmental debate on wind farm acceptance.

But if you have the time, the workshop, the site, and the passion, then you will build a windmill, and enjoy the hard-earned fruits. I hope this book helps. Be careful.

No free lunch

The wind is free, until the government manages to put a tax on it, and many people assume that wind energy will therefore be a bargain. If that were so, then we would see windmills everywhere, but of course 'there is no such thing as a free lunch'.

Wind is a very diffuse source of energy. To produce useful amounts of power, windmills need to be large; to work efficiently and reliably they need to be well engineered. So they are expensive. If you build your own, you can save most of the cost, but spend a great deal of your valuable time.

Battery depreciation

Power from small, stand-alone wind-electric systems using batteries is not likely to be cost-competitive with power bought from the national grid in the near future. Even the cost of the batteries can rule it out. Batteries may last about seven years before they are worn out. It has been calculated that just the cost of replacing the batteries can be roughly the same as the cost of buying the same amount of power as the system produces in this period, from the mains supply.

This comparison highlights the fact that battery-windpower is not likely to be viable in the city. (There are other reasons, such as low windspeeds, turbulence, and the fury of neighbours.) In remote places, the cost of installing and maintaining power lines may be greater than that of the windpower system, so windpower becomes a more economic and reliable source than the mains.

Pay less and get more from the scrapyard

I am a frequent visitor to my nearest non-ferrous metals dealer, where I collect cable, batteries, steel for welding, sheet aluminium, etc. Using scrap materials will not necessarily reduce the quality of the job. You can afford to buy something much heavier from the scrapyard than you could afford to buy new. For example, it was once possible to obtain 'scrap' batteries from telephone exchanges. I have used them for ten of their twenty year lifetime. If I had bought new batteries, I would only have been able to afford a small poor quality one. (However, use some discrimination. Most batteries are scrapped because they are useless, so check them carefully with a voltmeter before buying.)

Everything in the scrapyard is there for a reason, but very often the reason is modernisation. There may be nothing wrong with the 'scrap' you buy.

The environmental cost

Every source of power has an environmental price. Windpower is clean and renewable, but it does have some downsides. At least the pollution it causes is here and now, so 'what you see is what you get'!

Noise

There are two main kinds of noise which can arise: blade noise and mechanical noise. Blade noise is rarely a problem, as it sounds similar to wind in trees, or flowing streams, and is often masked out by these very sounds.

Mechanical noises can arise where there is vibration or hum from the generator or gearbox. These tonal noises can drive people crazy, especially if they are kept awake. Others (the owners) will enjoy the music of the windmill feeding power into the battery and sleep all the better!

Visual intrusion

Visual intrusion is even more subjective than noise. One person's sleek dream machine might be another's eyesore. A windmill will normally be attractive to the owner, especially if self-built. Neighbours may be willing to accept it, but tact and

———Table 1.1 Instant Power Outputs in Watts———

Windspeed:	2.2m/s 5 mph	4.5m/s 10 mph	10 m/s 22 mph	20m/s 44 mph
Blade diameter 1m	1	6	70	560
Blade diameter 2m	3	25	280	2,300
Blade diameter 3m	7	60	630	5,000
Blade diameter 4m	12	100	1,120	9,000

This table gives you an idea of how much power your windmill may produce. It assumes a modest power coefficient of 0.15. For example, a two metre diameter windmill in a ten metre per second wind might produce 280 watts. Do not be fooled by the apparent precision of the figure. In reality you may get between 200 and 400 watts, depending on what 'power coefficient' you can attain.

diplomacy are very important in gaining this acceptance.

How much power can you expect?

Power (in watts) is the rate of capture of energy, at any given instant. Table 1.1 shows how much power you can expect a windmill of a given size to produce in a given windspeed. The table assumes that your windmill catches 15% of the raw power in the wind. The percentage caught is known as the 'power coefficient' (or Cp) and we shall see later why it is such a small a part of the total.

The raw power in the wind depends on the density of air (about 1.2 kilograms per cubic metre), the speed of the wind and the size of the rotor. Windspeed is critical (as you can see from the table). Stronger winds carry a greater mass of air through the rotor per second and the kinetic energy per kilogram of air depends on the square of its speed, so the power in the wind will increase dramatically with windspeed.

The area swept out by the windmill's propeller, fan, sails, wings, turbine, blades depends on the square of the diameter. We call the windmill rotor blade assembly the rotor for short. Do not confuse this with the rotor of the generator.

At the back of this book there are windpower equations which you can use to calculate the power output of a windmill. Better still, use a spreadsheet to teach your computer to do the sums for you!

As you can see, the power in the wind varies enormously. There are only a few watts available in a light wind. It is not easy to design a machine which can convert this amount of power effectively, and yet survive the huge power surges which arise during gales.

The wind is always changing, and the power fluctuations can be extreme. We need to harvest it when it is there, and either store it for periods of calm, or use some other power source as a back-up. In the days of corn-grinding windmills, the millers kept a store of grain, and ground it as and when they could. Nowadays, small wind-electric systems use batteries, which absorb surplus power during windy weather, and keep the supply going during calm periods.

A quick guide to predicting energy capture

Energy captured in a given time is the average power multiplied by the hours. This depends at least as much on the site as on the machine itself.

Site conditions	Average windspeed
Trees and buildings	3 m/s (6 mph)
Open fields, with few hedges	4.5 m/s (10 mph)
Hilltops or coasts (open sites)	6 m/s (13 mph)

Average power output from a windmill is not the same as its instantaneous power output when windspeed is average. Again, there is an equation for this at the back of the book.

From Table 1.2 we can see that a two metre diameter windmill will give an average power output of about 51 watts where the average windspeed is 4.5 metres per second (10 mph). These are just ballpark figures: average output could in reality be anything from 30 to 80 watts.

What can you power from a windmill?

The average power from a windmill must be matched up to the average power needs of the user(s). The typical person (in Europe) has an average domestic electricity consumption (at home) equivalent to using 100 watts all the time. Sometimes they might use

——Table 1.2 Average Power Outputs in Watts——

Average windspeed:	3 m/s 7 mph	4.5m/s 10 mph	6 m/s 13 mph
Blade diameter 1m	4	13	30
Blade diameter 2m	15	51	121
Blade diameter 3m	34	115	272
Blade diameter 4m	60	204	483

many kilowatts, but at other times they hardly use anything.

So for a family of five an average power of 500 watts would be needed. But it is possible for a family of five to get by using under 100 watts if they use energy-efficiency lighting with care and avoid the use of electric heaters in low windspeed periods.

Efficiency: where does the energy go?

In Tables 1.1 and 1.2 we assumed that the windmill would catch 15% of the power in the wind. In reality, the power coefficient will depend on how much is lost at each stage of the energy conversion process. Some is even lost before it can begin.

Betz's theorem

Albert Betz (1926) is credited with figuring this out, so his name is always used to refer to this theory.

In order to extract power from the wind, it must be slowed down. To remove all the wind's power would involve bringing the air to a halt. However, this would cause a pile-up of stationary air at the windmill, preventing further wind from reaching it. The air must be allowed to escape with some speed, and hence with some kinetic energy (which is lost).

According to Betz, the best power coefficient we can hope for is 59.3%, but in practice this figure will be whittled down further by other losses described next.

Drag

The rotor blades convert the energy of the wind into shaft power. Later we discuss the advantages of using a few, slender blades which rotate fast, compared with many wide blades, rotating slowly. Fast moving blades will experience aerodynamic 'drag'. Drag holds the blades back, wasting some of the power they could be catching from the wind, so we need to make the blades as 'streamlined' as possible. Even the best designed 'airfoil section' blades will lose about 10% of the power they handle this way. Home built blades may lose a lot more.

Mechanical friction

There will also be friction losses in the bearings, brushes and any sort of mechanical drive, such as a gearbox or pulley system. These will only increase slightly with increasing speed. Therefore when the windmill is working hard, in a strong wind, the friction losses may be only a tiny percentage of the total power. But in light winds friction losses can make an enormous difference, especially in very small windmills, which have relatively low rotor torque.

Whether this is significant will depend on what is expected from your windmill. If it is your sole electricity supply, it will be crucial to have high efficiency in light winds and you should use direct drive from the rotor blades to the generator, with no gearing arrangements. Those who use the wind for supplementary heating only, for which light winds are of little use, may cut costs by using a geared motor, or a belt driven alternator, which will work well in a stiff breeze.

Copper losses

The next stage is to make electricity. This takes place in the coils (or windings) of the generator. Electric current suffers from its own kind of friction, which heats the wires.

This 'friction' is in proportion to the 'resistance' of the copper wires carrying the current (see windpower equations). You can reduce the resistance (and so the 'copper loss') by using thicker wires. This makes the generator heavier and more expensive, but it may be worth it.

The resistance of a copper wire increases with rising temperature. Copper losses heat the coils, which increases temperature, thereby increasing resistance and causing more copper loss. This vicious circle can lead to burn out in the worst case, and will certainly lower the efficiency of the machine, so it will be important to look at the cooling of the generator, in the overall design.

Copper losses increase with the square of current. When the generator is working at 'part load', in other words in light winds, losses in the main windings are very small. Some generators also have 'field coils' (see chapter five) carrying an almost constant current. These losses are rather like the mechanical losses discussed above. In light winds, they may consume all the power the blades can produce, leaving you with nothing.

Finally, do not forget about copper loss in the cable from the windmill. Where the cable is very long, it also needs to be very thick. If the cost of thick cable becomes ridiculous, then it is worth changing the system voltage. At higher voltages, less current will be needed to transmit the same amount of power. High voltage means much lower copper loss in cables, which is why it is used, in spite of the safety problems it may cause. A 12 volt system will lose 400 times as much power as a 240 volt system, when using the same cable.

Iron losses

Most generators also suffer from iron losses, which are described in detail in chapter four.

Rectifier losses

Very often, small windmills are built with permanent magnet alternators, which produce alternating current (a.c.). The power is then fed into a battery, for use as direct current (d.c.). A converter is required, which changes the a.c. into d.c. This is the 'rectifier'.

Modern rectifiers are simple, cheap, reliable semiconductor devices, based on silicon diodes. They work very well, but like everything in this world, they need their percentage. (One begins to wonder if there will be any power left at the end of all this!) In this case the rule is simple: each diode uses about 0.7 volts. In the course of passing through the rectifier, the current passes through two diodes in series, and about 1.4 volts are lost. In other words, to get 12 volts d.c. out, we need to put 13.4 volts in. This represents another energy loss, representing about 10% of the energy passing through the rectifier.

Again, changing to a higher voltage will reduce this loss. For example, in a 24 volt system the voltage lost in the rectifier will be

How the Losses Add Up

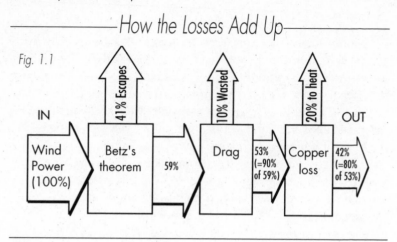

Fig. 1.1

IN

41% Escapes

10% Wasted

20% to heat

OUT

| Wind Power (100%) | Betz's theorem | 59% | Drag | 53% (=90% of 59%) | Copper loss | 42% (=80% of 53%) |

the same as in a 12 volt system (1.4 volts), but it is now less than 5% of the total.

How the losses add up (or rather multiply)

Each stage in the system passes on a percentage of the power it receives. We apply these percentages to each other in a row (Fig. 1.1) to get the overall power coefficient. It is fortunate that we are starting off with free energy! Of course, there are losses in any process. For example, every internal combustion engine always converts most of the energy in its fuel to heat, seldom recovered for any useful purpose.

Design basics
Matching the rotor to the generator

For a given size of rotor, it is tempting to use a very large generator, to make use of the high power in high winds. But, for a given size of generator it is tempting to use a very large rotor, so as to obtain full power in low winds.

A big generator with a small rotor will very seldom be operating at rated power, so it will be disappointing, especially if the generator's part-load efficiency is poor. A small generator with a large rotor will achieve full power in low winds, giving a more constant power supply. The drawbacks are that the larger rotor will:

• need a stronger tower (chapter 8);

• run at lower rpm (next section);
• require more control in high winds (chapter 6).

The usual compromise is to choose a generator which reaches full output in a windspeed around ten metres per second (10 m/s). See the first and the fourth columns of Table 1.1, or the first two columns of Table 1.3. (page 19).

It is also vital to match the rotational speed (rpm) of these two components, for which we need to understand their power/speed characteristics.

Tip speed ratio

The speed of the tip of one blade depends on the revolutions per minute (or rpm), and the rotor diameter. For example, the tip of a two metre diameter rotor, running at 500 rpm, travels at about 52 metres per second. This is over 100 mph! Operating tip speeds of up to 134 m/s (300 mph) are not unknown, but for the sake of a quiet life you should try to keep it below 80 m/s.

Tip speed ratio is the magic number which most concisely describes the rotor of a windmill. It is how many times faster than the windspeed the blade tip is designed to run. A windmill rotor does not simply have a best rotational speed (e.g. 600 rpm). Its optimum rpm will depend on the windspeed, the diameter and the tip speed ratio. (See windpower equations.)

The windmill rotor will do best at a particular tip speed ratio, but it will inevitably have to work over a range of speeds. The power coefficient 'Cp' will vary depending on tip speed ratio, for any particular rotor design. It will be best at the 'design' or 'rated' tip speed ratio, but acceptable over a range of speeds.

Figure 1.2 overleaf shows the power coefficient of a typical rotor designed to operate at a tip speed ratio of 7. A small shift in rpm or windspeed will not make much difference. If the rpm is too low, compared to the wind, then it will stall, and performance will drop. If there is no load on the rotor (perhaps because a wire has broken in the electrical circuit), the rotor will overspeed until it reaches a certain point, where it becomes so inefficient that it has no power to go faster. Most windmills are quite noisy and alarming at runaway tip speed.

In chapter three we shall look more closely at how to design a

————Power Coefficient and Tip Speed Ratio————

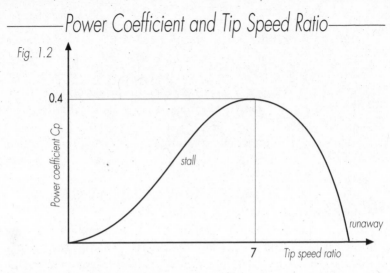

Fig. 1.2

windmill rotor to run at a particular tip speed ratio.

Generator characteristics

The rotor will accelerate until the load (generator) absorbs all the power it can produce. If the generator and the rotor are well matched, this will occur at the design tip speed ratio, and the maximum power will be extracted from the wind.

Generators also have their preferred speeds of operation. As we shall see later, the voltage produced by a generator varies with the speed of rotation. It will need to be run fast. If it is connected to a battery, then no power will come out of the generator until its output voltage exceeds the battery voltage.

The shaft speed (rpm) above which the generator delivers power is known as the cut-in speed. The speed required for full power output is known as the rated speed. These speeds need to correspond to the speeds at which the rotor 'likes' to run, in the corresponding windspeeds.

Finding the best rpm

Table 1.3 gives guidelines for matching speeds to generators. Choose the power you need in the first column. This is the rated output of the generator (and thus the windmill). The second column suggests a suitable rotor diameter, based on the assump-

—————Table 1.3 Rpm for Various Turbines + TSRs—————

Power (watts)	Diameter (metres)	TSR=4	TSR=6	TSR=8	TSR=10
10	0.4	2032	3047	4063	5079
50	0.8	909	1363	1817	2271
100	1.2	642	964	1285	1606
250	1.9	406	609	813	1016
500	2.7	287	431	575	718
1000	3.8	203	305	406	508
2000	5.3	144	215	287	359
5000	8.4	91	136	182	227

tions that your Cp is 15% and the rated windspeed is 10m/s. The remaining columns give figures for the generator speed required in rpm, for each of a series of possible rotor tip speed ratios.

Suppose you want 250 watts, using a tip speed ratio of six. Choose the fourth row. From the second column, read the suggested rotor diameter: 1.9 metres. What rpm must the generator operate at? Looking across we find that the fourth column has 609 rpm.

This brings you up against the hardest problem in small windmill design. It is impossible to find a generator with such a low rated speed. Generators work much better at high rpm. They are usually designed to give full output at between 1500 and 3000 rpm. Here are various ways around this problem, each with its own pros and cons which will unfold as you read this book:
- Gear up the speed between the rotor and the generator;
- Use a higher tip speed ratio;
- Work at a higher rated windspeed;
- Modify the generator to work at lower speed;
- Build a special, low speed generator.

You must also consider the cut-in speed. Ideally, the generator cut-in rpm should be about one third of its rated rpm. Keeping the rotor at its design tip speed ratio, this allows cut-in at about 3.3 m/s (assuming 10m/s rated windspeed). If the cut-in rpm is higher than half the rated rpm, then problems may be found in reaching this rpm in low windspeeds.

Summary

Windpower is fun but not free. There is a price to pay not only in pounds but also in your time and through an impact on other people's environment. You can use the tables in this chapter to select the size of machine needed. The tables take account of the losses for you by making some assumptions about the power coefficient. Speed-matching the rotor to the generator creates some dilemmas. Fast rotors are noisy, slow generators are heavy and gearing between the two wastes power.

Human life and happiness is of course more important than windpower, so the next chapter is about safety. After that we look at how to design and build windmills, from the rotors, through the electrics, to the tails and towers.

Chapter Two
Safety

For many people, experimenting with small windmills is stepping into the unknown, a real life adventure. If you were sailing a yacht, or wiring a 13 amp socket, there would be someone nearby to tell you the safe way. Far fewer people know about windmills. That puts a bigger responsibility on you to be safe.

Consult with experienced people where possible, but do not necessarily expect them to give the final word. A domestic installation electrician will probably be unfamiliar with variable-voltage 3-phase supplies, for example. Someone needs to know the risks, and that person is you.

Electrical safety

Electricity supplies present two main hazards: fire and shock. Both are covered thoroughly by the IEE wiring regulations, and many books are published to interpret these regulations. American readers should check the NEC code, which now includes sections specifically about renewable energy.

Protection against fire

In the last chapter we mentioned copper loss, whereby electric current flowing through a wire generates heat. When a wire is carrying too much current, it can become hot enough to melt the PVC insulation coating and set fire to the building.

——————Correct Use of Overcurrent Devices——————

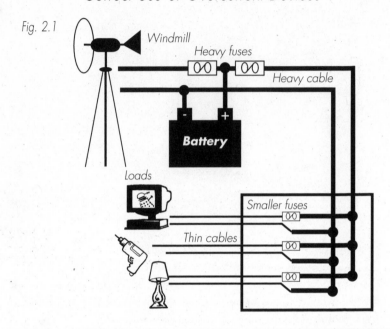

Fig. 2.1

Windmill

Heavy fuses

Heavy cable

Battery

Loads

Smaller fuses

Thin cables

Short circuits, fuses and MCBs

Excessive current may be due to overload, where too much power is being used from the circuit, or due to a 'short circuit'. A 'short' is the name given to a fault in which there is contact between the two wires from the supply (positive and negative, or live and neutral). A mains supply, or a battery, can deliver very high currents of thousands of amps when short circuited.

Whatever the cause, excessive currents need to be stopped before they start fires. Every circuit coming from a mains supply (or a battery) needs to be fitted with an 'overcurrent device', a fuse or a circuit breaker, which will break the circuit automatically if too much current flows (Fig. 2.1). Fuses are cheap to fit but cost money to replace. Miniature Circuit Breakers (MCBs) are increasingly popular, despite the extra cost. MCBs look just like switches, can be used to disconnect a circuit manually, and if they trip they are easy to reset. They are generally more sensitive than fuses, and therefore safer.

The heat produced depends on the size of the wire. If they use different sizes of wire, each circuit needs to be considered separately. The overcurrent device must be capable of carrying the current normally to be expected in the circuit and it must be designed to disconnect if the cable is overloaded or short circuited.

Bad connections and scorched walls

Cables are not the only fire hazards in an electrical system. A corroded connection will develop a high resistance to the flow of current before it fails completely. Normal current passing through this resistance heats it up, perhaps to the point where it can scorch the surroundings. Therefore:

- always mount connections on fireproof materials, not wood.
- prevent moisture from corroding the connections by keeping them clean and dry.

Heaters

Last but not least, there is a fire risk from incorrectly installed heaters. An electric heater needs good ventilation, and may need to be surrounded by fire-resistant materials for safety.

'Dump load' heaters are particularly hazardous. These exist for the purpose of disposing of surplus energy. They are normally controlled by an automatic control circuit, which operates without human supervision. If the dump load is rarely needed, it may come on unexpectedly after a long interval. It may by then have been covered up by old coats, or some other inflammable material.

Protection against shock

An electric shock is a current through the body. It happens because a person touches two different conductive objects, between which there is a voltage. There are several different ways to protect against the risk of shock.

Using extra-low-voltage

The simplest way to prevent shock is to use very low voltages such as 12 or 24 volts. Even if a person touches both terminals of the battery, there will be no sensation of shock (try it if you don't believe me). Voltages below 50V are termed 'extra low voltage' (ELV). If you keep them segregated from higher voltage circuits, these are relatively safe.

A word of warning about 'battery voltage'. The voltage rating of a windpower system is nominal, not exact. If the battery is disconnected, and the windmill is running fast, there will be much higher voltages coming from the windmill. Also, there are windmills which use transformers and high voltage transmission from the generator to the control box, in order to minimise cable loss. Never assume that the voltage from the windmill is too low to give you a shock.

Enclose it, fuse it and earth it

If you must use mains voltage, then it is essential to take precautions. The safest way to treat a mains voltage supply is to follow standard mains voltage wiring practice. This will make your system easier for others to understand. But remember that in practice your windpower supply may not behave just like the mains.

All live conductors must be inside a box, away from idle fingers.

By all means recycle cable from the scrap heap. But always check that the insulation (sheathing) on the cable is perfectly undamaged before you use it for mains voltage work.

The 'eebad' system

Mains supplies here in the UK are made safe by a system called 'earthed equipotential bonding, and automatic disconnection of supply'. 'Equipotential bonding' means connecting together every metal surface you are likely to touch. The bonding cable will 'short circuit' any dangerous voltage which may arise due to a fault. Bonding of electrical appliances is achieved by the 'earth' wire in the cable. Use your common sense to decide which other objects to bond together; where there is electricity in use, all exposed metal surfaces must be bonded. A dangerous voltage is unlikely between your knife and fork, unless you are eating inside your fusebox. But water and gas pipes need to be bonded to the earthing system.

Metal objects are not the only conductors you will make contact with. Planet earth is a conductor, so a voltage between the water tap and earth could give you a shock. Hence it says 'earthed equipotential bonding' in the recipe for safety. You should bond all

——————————*An Earth Fault Blowing a Fuse*——————————

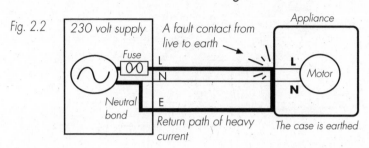

Fig. 2.2

'earth' wiring to one or more copper clad rods driven into the soil.

'Automatic disconnection of supply' is also required, so that when a fault occurs, it is quickly over. A 'fault' would be an untoward contact between a 'live part' (which should be insulated) and an exposed part (which is earthed). Such a contact will result in a dangerous situation and the supply must shut down.

In mains wiring, the automatic disconnection is often achieved using overcurrent devices. In this country, the neutral side of the supply is bonded to earth at the supply. Any contact between a live part and an earthed part is therefore a short circuit of the supply, causing massive overcurrents, which will operate the fuses or circuit breakers (Fig. 2.2).

Residual Current Devices (RCDs)

If the supply is a windmill or an inverter, then there may not be enough current forthcoming to blow or trip the device, even when the supply is shorted out directly. An overcurrent device is not a suitable 'automatic disconnect' for such a supply. A 'residual current device' (RCD) is needed. This is very sensitive, responding to a tiny leakage of current to 'earth' by tripping off the supply.

When you connect an RCD in your system, first check where the neutral is bonded to earth. There must never be more than one bond between neutral and earth, and it is usually made at the supply. Where a number of alternative supplies are used, neutral should be bonded at the distribution board. The bond between neutral and earth must be on the supply side of the RCD, or the RCD will not 'see' the fault current at all (Fig. 2.3 overleaf).

Correct RCD Positioning

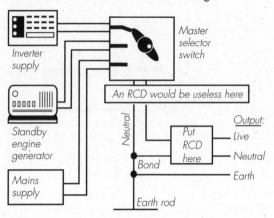

Fig. 2.3 The RCD must be downstream of the neutral bond.

Protective devices summarised

To prevent fires, fit an MCB at the supply end of the live (or positive) side of every cable in the system. The current-carrying capacity of the cable must be at least as large as the MCB's rating, which must be greater than the intended load current to be used.

To avoid shock hazard, check that the neutral is bonded to earth at the supply, and fit an RCD.

Battery hazards

I hate batteries. They are the worst feature of stand-alone windpower systems. I wish we did not need batteries, but they are essential where wind energy is to be the primary source for your electricity. They combine several dangerous features.

Firstly, they are heavy enough to damage your spine when you shift them around.

Secondly, they are full of corrosive sulphuric acid, which attacks your clothing and your skin. It is especially hazardous to the eyes. Clean up spills with an alkaline solution (e.g. washing soda, some of which you should always have handy). In the event of skin (or eye) contact, wash with plenty of water. (Wash your overalls promptly too, unless you want the acid to eat holes in them.)

Thirdly, they will kick up a terrific spark and could give you a nasty burn if you short circuit them, with a spanner for example. Remove all jewellery when working with batteries. Always fit a fuse or circuit breaker to protect the wiring. Do not let the overcurrent device touch the battery or it will corrode.

Lastly, they give off explosive gases, which can be ignited by a spark, so keep batteries in a ventilated enclosure. A small vent at the highest point, leading to the outside, is all you need — the hydrogen gas rises rapidly. Never create sparks around the vents on top of a battery case. Sparks are a common cause of explosions within the battery. The trapped gases explode, and blow off the top, scattering acid around.

Dispose of batteries responsibly. Lead is toxic, acid is toxic, and both need to be recycled, not dumped. Scrap merchants will sometimes pay to take them off your hands. Your local authority will be able to help you dispose of them safely too.

Other responsibilities
Working with revolving machines

Windmill blades spin at high speeds. The blades need to be out of reach of passers-by and children.

When you complete your first set of rotor blades, do not be tempted to test them out by holding them up to the wind in some makeshift way, in your excitement. Once the rotor starts, you will not be able to stop it. Gyroscopic forces will probably twist it out of your grip, and someone is likely to get hurt.

Exposed belt drives, shafts and suchlike are dangerous. When experimenting with generators and drive arrangements at ground level (for instance on a lathe) you must treat them with great respect. Wear no loose clothing, and keep your hair out of the machinery!

Here is a story told to me by Mick Sagrillo, a windmill guru of some standing in the USA:

"I was up in Alaska on the Yukon river, 35 miles by boat from the nearest telephone. I was working with someone on his Jacobs, bench testing it on the ground with a battery. It was about 10 pm, and, of course, we had worked way longer than we should have. At the time, I wore a ponytail half way down my back, but tucked up under a cap.

"Somehow, the slow spinning generator shaft got a hold of my hair,

Working at Heights

Fitting a windmill rotor at the top tower using pulleys.

and proceeded to slowly tear it out of my head. All of it. Once I got home,
I wound up having surgery on my spine. It was quite an experience. I still
cringe at the sound of Velcro!!! Anyway, I was lucky because I was on the
ground, and I survived. And fortunately, my hair grew back. Needless to
say, I don't wear my hair long any more!"

Working at heights (avoid if at all possible!)

If you cannot avoid working at a height, then be sure:
- that the tower cannot fall down;
- to tie yourself on while working;
- that no one walks beneath while you are working, in case you
  drop a spanner.

Never climb unless you are confident, relaxed and sure that you
have planned your every move. Better still, don't do it at all.

The windpower industry in the USA has suffered several fatalities over the years, all related to falling from towers. In Europe it
is more normal to raise the windmill and tower as a unit, after
assembly. Whilst this also has its dangers, I know of no fatalities.

Lifting operations

Even when you never leave the ground, tower erection needs to be done with care and forethought. The leverages are surprisingly unpredictable, and there are often several different people working to several different plans at once.

The safest situation is where there are as few people as possible involved, one person is unmistakably in charge, and everyone stands well clear of the fall zone (where the tower might land if things go wrong). Lifting operations should be slow and deliberate. Sudden jerks and surges put a much greater load on your gear. All tackle must be stronger than required to take the load, by a good safety factor (say five times stronger).

Bits drop off and towers fall down...

Safety comes first and last in the design of windpower systems. Build your tower so that it can never fall down. Position it on the assumption that it surely will fall down, or at least shed heavy metal bits just at the moment when someone walks beneath.

Don't get too depressed. Safety consciousness doesn't take long. Better to be depressed before than after, when it's too late.

Chapter Three
Rotor Design

This chapter is about designing rotor blades for windmills. We start with the theory of how the blades do their job, and the next chapter describes the practical nitty gritty of how to make them.

Betz Revisited

There is a limited amount of power in the wind. We extract this power by decelerating it (slowing it down). According to Betz (chapter one), there is an optimum deceleration of the wind (down to one third of its upstream velocity) which extracts the most power (59.3%). More deceleration will simply divert wind away from the rotor.

The 'braking forces' which decelerate the wind are equal to the thrust force which the wind applies to the rotor (by Newton's Third Law). The essential trick of rotor design is to create the thrust force which produces Betz's optimum deceleration and no more.

How many blades?

A fundamental decision to be made in any rotor design is the number of blades to use.

Many people intuitively feel that more blades will produce more power. On the face of it, since each blade contributes to the power which the windmill produces, this does seem logical. The flaw in this logic is the underlying assumption that there is more power to be had. If two blades is the right number, then there is no

The Forces on a Windmill

Fig. 3.1

The windmill is pushing back at the wind to slow it down.

The wind is pushing the windmill backwards too (it's the same force!).

---

point in adding a third one, which will actually only get in the way.

So why do we see some windmills with many blades, while others have only two or three? The answer lies in the different jobs the windmills have to do, which require different speeds of operation.

The faster the blade sweeps across the wind, the larger the thrust force. In fact the force increases with the square of the blade's speed, so the effect of doubling the rotor rpm is that the force increases fourfold. There is an optimum force to develop, for maximum power extraction (following Betz). Doubling the speed of a rotor increases the effect of each blade on the wind fourfold, so we only need one quarter as many blades (Fig. 3.2, over).

Torque and speed

Mechanical power has two ingredients: force and speed. Torque is the technical expression for 'turning force'. Pumps require lots of torque, especially when starting from rest. Generators require lots of speed. They use the same amount of power, but in different ways. The power produced by the rotor is the product of both torque and rpm.

Table 3.1 (page 33) gives the typical choices for tip speed ratio and blade number for pumps and generators.

—Blade Type and shaft speed—

Fig. 3.2   **Generator: High Speed**   **Pump: Low Speed**

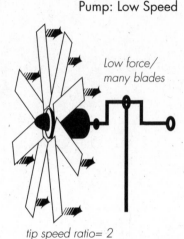

High force/few blades   Low force/ many blades

tip speed ratio= 4   tip speed ratio= 2

Two blades or three?

The number of blades we use is largely dictated by the tip speed ratio at which we wish to operate. There is some leeway however, as you can see from Table 3.1. Three narrow blades will have a similar performance to two wide ones.

What difference does it make, then? Two blades rattle more than three. This is partly because there is a difference in windspeed between the top and bottom halves of the windmill's swept area (Fig. 3.3) known as 'wind shear'. The force on the bottom blade will be less than the force on the top one. Where there are only two blades, this leads to a pulsating imbalance. The difference is worst in the vertical position, negligible in the horizontal position, and very complicated during the diagonal stages. Three bladed rotors do not suffer from this problem.

Another kind of wobble occurs when the windmill 'yaws'. Yawing is the term for 'swivelling' on the tower top. Two bladed rotors undergo a judder in yaw. Consider the effect of holding a long straight object, like a broomstick, when you try to make a sudden turn. If you are holding the broomstick up vertically, you can spin around on the spot with no trouble at all. When held

─────── Table 3.1 Turbine Types ───────

Tip Speed Ratio	No. of Blades	Functions
1	6 - 20	Slow pumps
2	4 - 12	Faster pumps
3	3 - 6	Dutch 4-bladed
4	2 - 4	Slow generators
5 - 8	2 - 3	Generators
8 - 15	1 - 2	Fastest possible

─────── The Effect of Windshear ───────

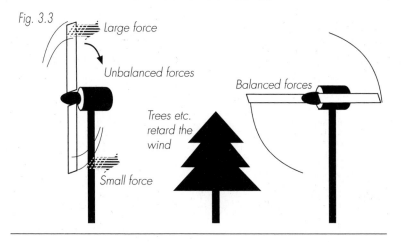

Fig. 3.3

Large force

Unbalanced forces

Balanced forces

Trees etc. retard the wind

Small force

horizontally, it has much greater inertia, making your turn much more ponderous. A spinning windmill rotor is moving from vertical to horizontal many times each second, so the yaw motion is very jerky (Fig. 3.4, over).

Using lift and drag

It's time to look more closely at how the windmill blades interact with the wind to produce the forces which drive the machine around. Any object finding itself in an airstream will get pushed about. The force on the object may be aligned with the air flow, but it is more likely to be skew. Asymmetrical objects create

Wobbles during Yaw

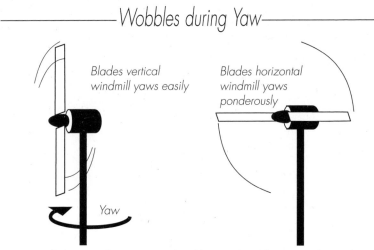

Blades vertical
windmill yaws easily

Blades horizontal
windmill yaws
ponderously

Yaw

Fig. 3.4 Blades rotate between vertical and horizontal rapidly, causing the machine to judder when yawing.

Lift and Drag

Fig. 3.5

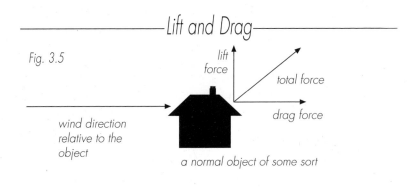

lift
force

total force

drag force

wind direction
relative to the
object

a normal object of some sort

Fig. 3.5 The components of aerodynamic force.

asymmetrical forces. To help in our calculations, we split the force up into two components, acting in different directions, called lift and drag (Fig. 3.5).

• Lift is at right angles to the flow.
• Drag is parallel to the flow.

Different types of windmill use different components of the force. The earliest known windmills were of the 'vertical axis' variety, probably derived from 'ox mills' powered by draught

Early Windmills

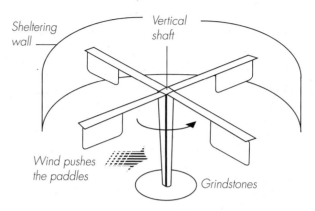

Fig. 3.6 Early windmills were of this type, and known as 'drag' machines.

animals walking in circles. The drive shaft was vertical, turning a grindstone in most cases. The blades were crude 'paddles' made from wood or matting, which caught the wind and were pushed around by it (Fig. 3.6).

A sheltering wall was used to protect the paddles on one side of the windmill, so that the wind pushed the rotor around. Alternatively, the paddles were designed to yield when the wind was from one side, but to catch the wind from the other. Either way, it was the drag force of wind on paddle that pushed the windmill around.

Windmills like this are known as 'drag machines' because they operate through the drag forces of the wind. Apart from being easy to understand, drag machines have few advantages. Half of the rotor is not doing anything, or is even holding the machine back. So the power coefficient is low. The speed is also limited. The tip cannot travel faster than the wind pushing it! Modern windmills operate on lift forces, and can therefore be called 'lift machines'.

What wind does the blade actually 'see'?

Imagine yourself on the blade tip of a typical modern windmill (Fig 3.7). You can imagine the effect which the downward movement of the blade tip has on the wind it experiences. If the rotor were revolving on a calm day, you would feel a 'headwind'

—A Modern Windmill—

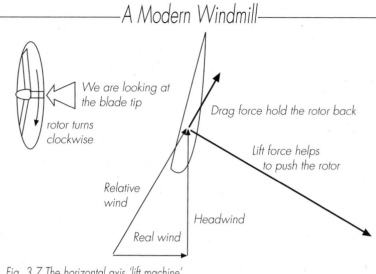

We are looking at the blade tip

rotor turns clockwise

Drag force hold the rotor back

Lift force helps to push the rotor

Relative wind

Headwind

Real wind

Fig. 3.7 The horizontal axis 'lift machine'.

equal to the speed of the blade. When there is also a wind from the side, the headwind adds to it, creating a resultant wind direction at an angle. We can call the wind experienced by the blade the 'relative wind'.

This kind of windmill is harder to understand, but has some big advantages. The tip can travel faster than the wind. The main difficulty arises from the drag force, which holds the blade back. The faster it goes, the more the drag force is rotated to hold it back (pointing upwards). To achieve a good power coefficient, we need to create the optimum lift, while minimising drag. In other words, we need to minimise the drag/lift ratio (drag divided by the lift).

It helps to use a more aerodynamic shape, for example a wing section, like the NACA 4412 section, in Figure 3.9.

The angle of attack is the angle between the chord line and the direction of the approaching wind, as seen by the wing section. There are graphs in data books which show how lift varies for common wing sections, at different angles of attack. Figure 3.8 shows the graph for the NACA 4412.

Notice that the lift coefficient increases as the angle of attack increases until we reach a point, known as 'stall', where the airflow over the back of the section separates off, and a zone of turbulence appears. A stalled wing has low lift, and very high drag.

A Typical Airfoil

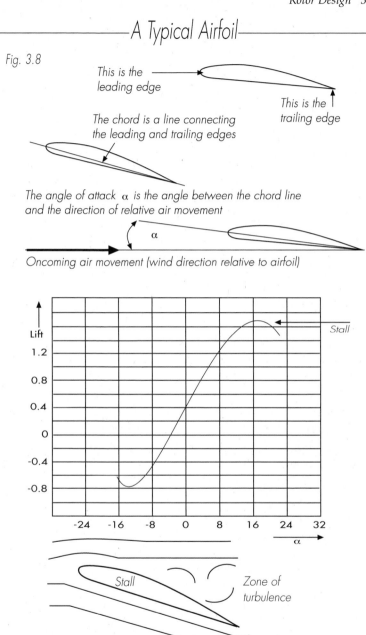

Fig. 3.8

This is the leading edge

This is the trailing edge

The chord is a line connecting the leading and trailing edges

The angle of attack α is the angle between the chord line and the direction of relative air movement

α

Oncoming air movement (wind direction relative to airfoil)

Lift
1.2
0.8
0.4
0
-0.4
-0.8

-24 -16 -8 0 8 16 24 32

α

Stall

Stall

Zone of turbulence

Fig. 3.9 Lift versus angle of attack for the NACA 4412 (Re=10^7).

Rotor Dimensions and Stations

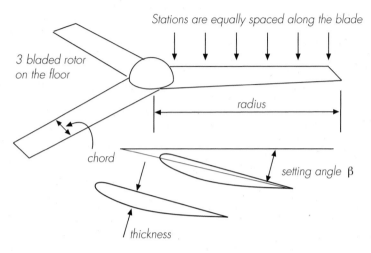

Fig. 3.10 Rotor dimensions are specified at stations along the blade. The 'chord' is the depth of the blade.

Wind tunnel studies show that the drag to lift ratio is not a constant factor: it varies as we tilt the wing section. The best ratio is usually at an angle of attack of around 4°.

This data in table 3.2 is interesting because it shows that there is not a huge variation in lift between these different shapes, but there is in the drag/lift ratio. Streamlined shapes, like the NACA section have much less drag than crude shapes. The position of the tube on the back of the blade makes the fourth shape particularly bad.

Blade design

Designing a blade, we specify the shape at each of a series of 'stations' along its length (Fig. 3.9).

At each station, the following data are given:
- radius
- setting angle
- chord
- thickness.

Radius

This is simply the distance from the centre of the rotor to the station.

Table 3.2 Data for some Simple Sections

Section		Drag/ Lift Ratio	Angle α	Lift Coefficient C/L
Flat plate	————	0.1	5°	0.8
Curved plate (10% curvature)		0.02	3°	1.25
Curved plate with tube concave side		0.03	4°	1.1
Curved plate with tube convex side		0.2	14°	1.25
Airfoil NACA 4412		0.01	4°	0.8

A list of sections that could be used, with their best drag/lift ratios.

Setting angle

Sometimes referred to as 'pitch', the setting angle (β) is the angle between the chord line and the plane of rotation of the windmill rotor. Here is the procedure for finding the optimum angle:

1. Decide what angle of attack we want to operate at (usually 4°) for minimum drag/lift.

2. Find the direction of the relative wind striking the leading edge of the blade at each station (Fig. 3.10). This will be the sum of two velocities: the wind velocity through the rotor, and the 'headwind' velocity caused by the rotation of the rotor itself. We can call this the flow angle Φ.

3. The setting angle is the difference between the flow angle and the angle of attack.

The flow angle (and so the setting angle too) depends on how far out along the blade you go (Fig. 3.11 overleaf). Near the root, the wind comes in almost square-on to the rotor, and the setting angle needs to be large. Out at the tip, the headwind is much larger, and so the direction of the relative wind is rotated, and the setting angle should be much smaller. Table 3.3 (overleaf) suggests suitable setting angles (in degrees) for five equally spaced stations and four possible tip speed ratios.

Adjusting the Setting Angle

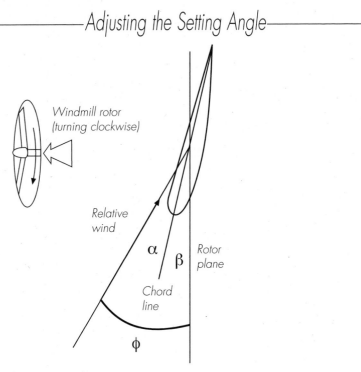

The setting angle, β, equals the flow angle, φ, minus the angle of attack, α.

Fig. 3.11 The setting angle is adjusted for the best angle of attack.

Blade width or chord

The next stage of the blade design process is to specify the chord width of the blade at each station along the span. To calculate the chord from theory, we equate the aerodynamic thrust (from lift calculations) with the force required for the Betz change of momentum (Newton's Laws) and eventually arrive at a workable formula. The simple formula in the windpower equations (at the back of the book) works reliably for fast running rotors.

Table 3.4 gives examples of chord widths for a selection of tip speed ratios. In each case, the number of blades 'B' is chosen for you, and the chord width is given as a percentage of diameter.

Flow Angle at Root and Tip

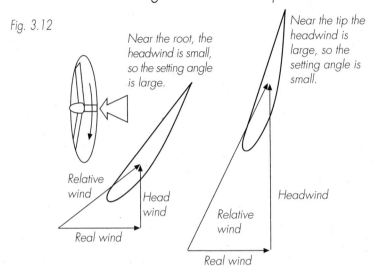

Fig. 3.12

Near the root, the headwind is small, so the setting angle is large.

Near the tip the headwind is large, so the setting angle is small.

Relative wind

Head wind

Real wind

Relative wind

Real wind

Headwind

Table 3.3 Setting Angles

at a series of stations (assuming that α = 4°)

	Setting angles (degrees) for tip speed ratio			
Station	4	6	8	10
1	36	25	19	14
2	19	11	8	5
3	11	6	4	2
4	8	4	2	1
5	5	2	1	0

Untwisted blades

The ideal chord width (like the ideal setting angle) goes to extremes near the root. However, there is only a little loss of performance if you use a simple, rectangular (parallel sided) untwisted blade. The area of wind swept by the inner part of the blade is relatively small. You could use the angle and chord specified for

———Table 3.4 Chord Width as a % of Diameter———

Tip Speed Ratio:	4	6	8	10
Number of blades:	3	3	2	2
Station 1	21.4	12.3	11.6	7.8
2	15.4	7.5	6.5	4.2
3	11.2	5.2	4.4	2.9
4	8.7	4.0	3.4	2.2
5	7.1	3.2	2.7	1.7

Chord width as a percentage of the diameter of the rotor. The columns give options for tip speed ratio and number of blades as illustrated in Figure 3.13. Station 1 is nearest the hub, station 5 nearest the tip, so the chord width gets thinner nearer the tip.

the fourth station over the whole blade span. To some extent, the narrow chord and flat setting of the inner part of the blade compensate for each other.

Why bother making twisted, tapered blades? Here are three good reasons:

- The efficiency is slightly improved;
- Tapered blades are stronger. The largest bending stresses are at the root of the blade, and a heavily tapered blade is much less likely to snap off under arduous conditions, or as a result of fatigue, than is a straight blade;
- Tapered blades are better at starting. The wider, coarser root gives slightly better torque. Every little helps!

Blade thickness

Thin sections have better drag/lift ratios, so they should be used where possible, for best performance. Near the root, where the speed ratio is low, the drag/lift is not so important, but the strength is, so a thick section is more appropriate.

If only one section is being used for the whole blade, then NACA 4415 (with thickness 15% of chord) is a good compromise.

Upwind, downwind or vertical axis

There are various different orientations of rotor you could use (Fig. 3.5). Most windmills are what are known as 'horizontal axis

Blade Shapes

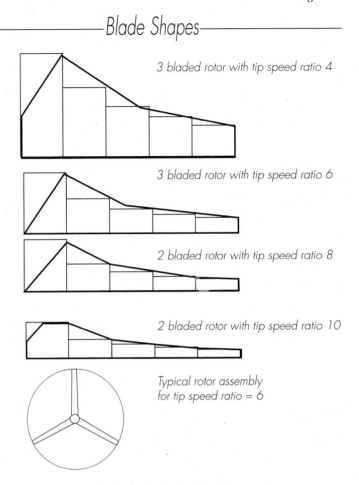

3 bladed rotor with tip speed ratio 4

3 bladed rotor with tip speed ratio 6

2 bladed rotor with tip speed ratio 8

2 bladed rotor with tip speed ratio 10

Typical rotor assembly for tip speed ratio = 6

Fig. 3.13 These shapes are for the chord widths in the table opposite.

wind turbines' or 'HAWTs'. Among the HAWTs, there are the upwind and the downwind varieties, depending on whether the rotor is upwind or downwind of the tower. There is another kind known as 'vertical axis wind turbines' or 'VAWTs'.

Downwind rotors are often 'free coning' (Fig. 3.14), meaning that the blades are mounted on hinges, or are flexible. Wind thrust throws them downwind, but centrifugal forces throw them outward. The blades adopt a particular coning angle, which depends on wind strength and rotor speed. The purpose of this

—A Downwind Windmill—

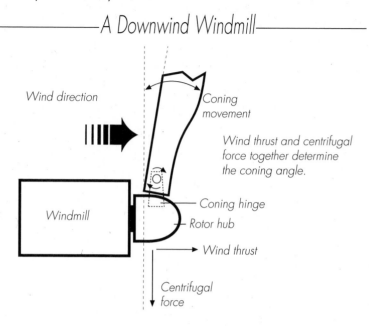

Wind direction

Coning movement

Wind thrust and centrifugal force together determine the coning angle.

Coning hinge

Rotor hub

Windmill

Wind thrust

Centrifugal force

Fig. 3.14 This design has coning hinges on the blades.

'freedom' is to remove all the bending stresses at the blade root. Being downwind of the tower, the rotor passes through the tower wake, so the blades lose most of the wind thrust force once per revolution. This continuous waggling can cause the blades to snap off at the roots.

Coning hinges are also used on single bladed rotors, together with a counter-weight. Coning freedom is not such a good idea for upwind rotors, as sudden gusts can push them back into the tower.

Another advantage which is claimed for the downwind rotor is that it can be 'free yawing'. It is not easy to put a tail on one of these machines, nor is it necessary, if all is well. Under certain conditions of tip speed ratio, however, a downwind windmill may decide to run upwind. This is very difficult to explain, but impressive to watch, when it happens.

Vertical axis machines

We have already looked at one vertical axis wind machine in this chapter: the drag machine which took over from oxen. A similar design is still used for anemometers (instruments which

Different Rotor Orientations

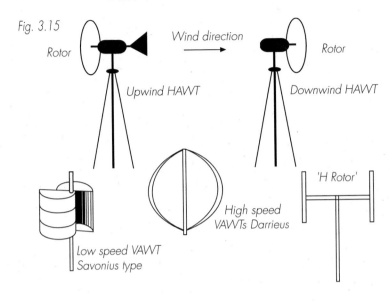

Fig. 3.15

Rotor

Wind direction

Rotor

Upwind HAWT

Downwind HAWT

'H Rotor'

High speed
VAWTs Darrieus

Low speed VAWT
Savonius type

measure windspeed), because of its very constant tip speed ratio. As a prime mover it is not so competitive, because it is very inefficient and slow.

The Savonius or 'S' rotor type of VAWT (Fig. 3.15) is very popular among DIY enthusiasts. It has plenty of starting torque, and is excellent for rotating signs, ventilator rotors, etc., but has very poor efficiency, and is rather slow for electricity production. Maximum power coefficient (Cp) for the Savonius rotor is about 0.15, whereas HAWT rotors can usually exceed 0.3.

There are faster, high efficiency 'lift' machines among the vertical axis types as well. These windmills work by lift, like HAWTs do, rather than drag.

VAWTs which rely on lift forces like this need to operate at low drag/lift ratio. To achieve this, the angle of attack must be small. The rotor has to turn fast, so that the large headwind keeps the angle of attack below stall. This means a high tip speed ratio. The high tip speed ratio dictates few, narrow blades (low solidity).

Forces on VAWTs

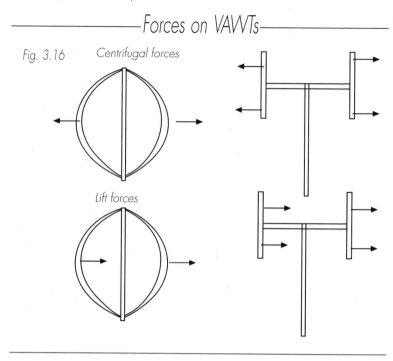

Fig. 3.16   Centrifugal forces

Lift forces

Fatigue stresses on high speed VAWTs

The narrow, vertical, airfoil-section blades of high speed VAWTs are subject to the bending effects of centrifugal and lift forces, which both act horizontally (Fig. 3.16).

The 'H' rotor suffers from the bending effects of both forces on the vertical blades. In the Darrieus or 'egg-beater' rotor, the bending effects of centrifugal force are defeated by curving the blades into a catenary shape. This is the same shape as a rope would take up, and the centrifugal force creates only tension, without bending. The lift forces still push the blades to and fro horizontally, however. They impose heavy cyclical loads on the blades, which consequently have a reputation for fatigue failures.

HAWTs also experience strong, horizontal lift forces on the blades, but they are less damaging because they are steady and not reversing all the time. Very large HAWTs can run into problems with reversing gravity forces, but this is not likely to be a problem for readers of this book!

—*A Shrouded Rotor*—

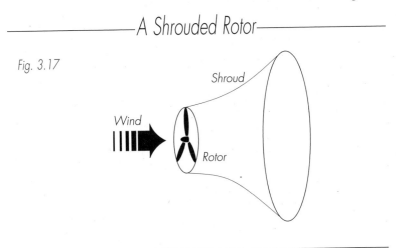

Fig. 3.17

Shrouded rotors

It is possible to fit a shroud to a windmill rotor (Fig. 3.17), which will 'funnel the wind' into the rotor. This brings great advantages, because it results in higher windspeeds through the rotor, giving more power and higher rpm. On paper it seems an obviously good idea. In practice, the construction of a suitably shaped shroud, which will not be damaged by high winds, is quite a challenge.

Shrouded rotors are sometimes included in the design of buildings. Where there is a high degree of exposure, such as with tower blocks or lighthouses, and where the walls can be shaped so as to accelerate the prevailing wind through apertures, the idea has some promise. A purpose built shroud, on the other hand, is likely to be expensive, ugly, and vulnerable to storms.

Conclusion

The options for clever rotor configurations are limited only by the extent of human folly, and they have all been tried! Making windmills is hard enough without adding unnecessary obstacles, so you are advised to use the most successful design, the three bladed upwind HAWT with a tip speed ratio of around six. This is the sort of rotor we shall look at in chapter four.

Chapter Four
Blade Making

This chapter explains how to make rotor blades. But before you start, make sure that the design speed of the rotor will be correct for the operating speed of your generator.

Table 1.3 will help you to choose a suitable diameter to produce your required power. Choose a tip speed ratio in accordance with your proposed generator speed and other design priorities. Tables 3.3 and 3.4 give details of the shapes of rotors with various tip speed ratios. Or you can set up a spreadsheet using the equations in the back of the book and design your own blades.

A word of warning

The blades are the most highly stressed component of the windmill. If a blade breaks (as has been known to happen to most types of windmill), it may fly some distance, and could cause serious damage or injury. If you plan to site your windmill within 100 metres of any public areas, paths or inhabited buildings, then you should first satisfy yourself that the blades are sufficiently strong.

Gyroscopic forces are the worst threat to small windmill rotors on turbulent sites. A fast running rotor, yawing to face changing winds, will suffer gyroscopic moments which alternately flap the blades back and forth once per revolution. Unless the blades are very securely attached they will fly off.

Blade weight

A heavy rotor will be harder to start than a light one, but once it is running it will be less likely to stop. The extra inertia of a heavier blade does not actually consume power, but it may prejudice the ability to start up during brief gusts of above-average windspeed.

Heavier blades suffer from worse centrifugal and gyroscopic forces, so there is little value in making a blade stronger by making it heavier. Strength-to-weight ratio is the important thing. For minimum stress, the blade should be light at the tip and strong near the root.

Blade materials

Wood is probably the best choice for one-off construction of windmill blades, because it is light, strong, workable, and has good fatigue properties. Metal (especially aluminium) is prone to fatigue. Steel is heavy. Plastic is suitable for blade manufacture and in fact glass reinforced polyester resin is the commonest material in use. But you need a mould which is as much work to carve as a wooden blade. Thick sheet plastic can be wrapped around a central spar to produce an acceptable section.

This chapter focuses on carving blades from solid wood. Lightweight softwood is the most suitable. Select close-grained timber which is free from knots if possible. Often you can plan your cutting out so as to marginalize the knotty areas. Avoid 'sapwood', because it has quite a different density from 'heartwood'. Sapwood, the outer layer of the tree trunk, is often the most knot-free part, unfortunately. You can recognise it by the lighter colour, relative to the heartwood. If only one blade contains sapwood, then the rotor will be difficult to balance. Also, the sap tends to migrate outward under centrifugal force, causing further balance problems, and spoiling the paintwork.

Blade timber must be well seasoned (dry). Oregon Pine is ideal, but expensive. Old church pews may be a good secondhand source. Manufacturers in the US use Bass, but this is not so readily available in the UK. There are some lovely rainforest hardwoods (e.g. Meranti) with a low density and a lovely texture, but these may not all be produced by sustainable methods; you would need to check with your supplier.

Some people advocate laminated timber for blades. There is something to be said for glueing pieces together, especially if the quality of the original timber is poor, as the defects will be less critical this way. Most of the grain should run along the blade, as the stress is all this way. Use good quality waterproof glue, such as epoxy.

Plywood sheathing can be used to fabricate hollow blades, for larger rotors where the weight of solid wooden blades is a problem.

How to carve a set of rotor blades

Here is a detailed, step-by-step description of the process for carving a set of blades for a three bladed rotor with diameter of 2.3 metres, and tip speed ratio of around 5.5. You can adapt the technique for any other tip speed ratio by adjusting the dimensions.

Tools

You will need the following: a hand saw (and optionally jigsaw or bandsaw), wood chisel (and mallet), plane, spokeshave, draw knife (recommended, if you can find one), callipers, compasses, square, tape measure, ruler, pencil, spirit level, drill.

Keep your tools very sharp, using an oilstone. The angle of the edge of the tool is quite critical. Always start by honing the tool, and then work up to the edge. Go easy on the edge itself, or you will actually make it blunter. Finish by removing any rag from the edge, using swift, light longitudinal strokes, or stropping the edge on leather or wood.

Always work with the grain of the wood, to prevent it splintering. Grip the workpiece firmly to a bench with a G clamp. If a tool judders or sticks, try sliding it sideways as you cut. A slicing motion like this gives more control.

Materials

3 pieces of wood, 150 mm by 50 mm, by 1150 mm long.
2 plywood discs, 12 mm thick, 300 mm in diameter, exterior or marine grade.

──Table 4.1 Summary of Finished Dimensions──

Station	Width	'Drop' (step 3)	Thickness
1	145	50	25
2	131	33	20
3	117	17	18
4	104	10	15
5	90	5	11

Mounting bolts to suit your hub.

48 galvanised woodscrews, size 40x4 mm countersunk.

If you have access to a thicknesser machine, you should start by passing the three pieces of wood through, to remove any warp and give a smooth straight finish. Do not worry if you lose a few millimeters off the overall dimensions, provided that all three are identical.

Step 1. The stations

Mark out the stations on the pieces of wood (Fig. 4.1), equally spaced at intervals of 230 mm. Draw a line right around the piece, using a square. The left hand end is the root of the blade, which will be at the centre of the rotor. The fifth station is the tip.

Step 2. Taper the blade

Measure the widths in millimetres (Table 4.1) from the edge which is nearest to you, and mark dots. Join the dots with a line. If there are any knots in the piece, turn the wood over so that they are in the triangular piece at the back which you will remove. You can use a bandsaw, or cross-cut the waste and split it out in sections using a chisel. Plane the newly cut surface smooth, straight and square.

Try to visualise the shape of the finished blade (Fig. 4.1). The tip moves clockwise, viewed from upwind, so the leading edge is the one nearest to you. The front (or windward) face is uppermost now. It should be perfectly flat (untwisted) at this stage. If not, then plane it flat. Check for twist with a spirit level, laid across the piece at each station in turn.

Carving the Blades

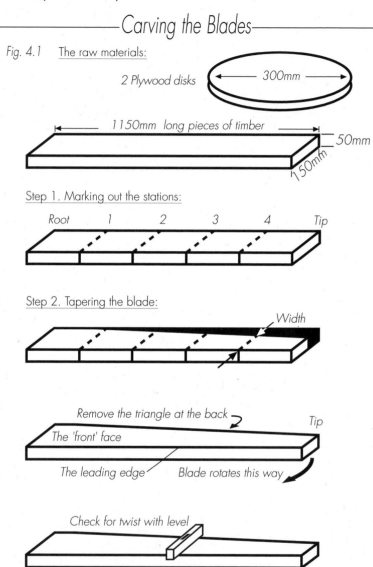

Fig. 4.1   The raw materials:

2 Plywood disks

⟵ 300mm ⟶

⟵ 1150mm long pieces of timber ⟶

50mm

150mm

Step 1. Marking out the stations:

Root   1   2   3   4   Tip

Step 2. Tapering the blade:

Width

Remove the triangle at the back

The 'front' face

Tip

The leading edge   Blade rotates this way

Check for twist with level

Step 3. Carving the twist

The next view (Fig. 4.2) shows the piece turned around, so that the leading edge is at the back, and the tip is on the left. At each station, draw a vertical line on the newly cut face, square to the front face.

Carving the Twist

Fig. 4.2

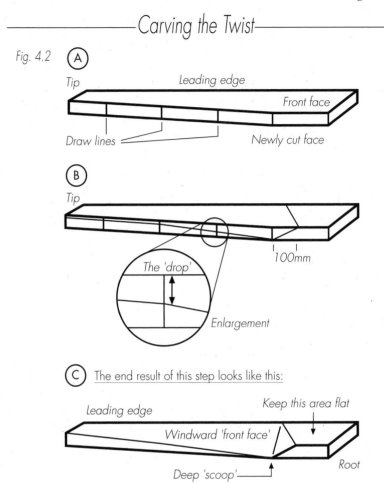

Mark a point on each line, a certain distance down from the front face. We call this distance the 'drop' (Table 4.1). It determines the setting angle at that station. Join the dots to draw the line of the trailing edge of the blade. Carve away all the wood above the trailing edge (pencil line). The windward face should end up so flat that when you lay a ruler edge across the blade between the leading and trailing edges, it will not rock.

At the root, the pencil line must rise again to the uncut face in a steep ramp 100 mm long. The root is to be left uncut, for assembly between the hub disks. (Note: with larger blades, it is easiest to use

Cutting out a Rotor Blade

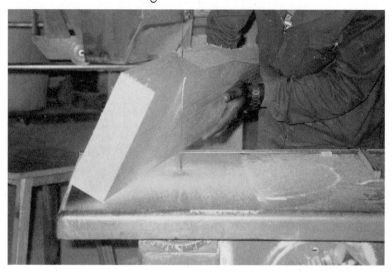

Carving the Thickness

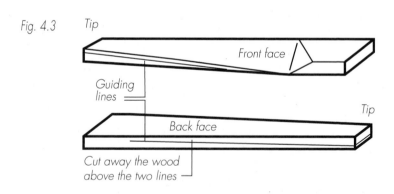

Fig. 4.3   Tip
Front face
Guiding lines
Tip
Back face
Cut away the wood above the two lines

---

a bandsaw to remove much of this deep 'scoop' near the root — see the above photograph).

Step 4. Carving the thickness

You now have a slightly tapered piece of wood, with a twisted face hollowed out of the 'front'. The next step is to remove wood

————————Two Tools You Can Make Yourself————————

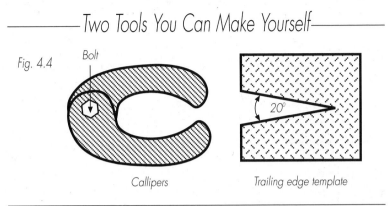

Fig. 4.4

Bolt

20°

Callipers   Trailing edge template

from the back of the piece, so that it is the correct thickness at each station (see Table 4.1).

Lay the piece of wood with leading edge uppermost. At each station, make a mark at the correct (thickness) distance from the leading edge. Join the dots with a line. Turn it over and do the same relative to the trailing edge. Now you have two lines (Fig. 4.3) to guide you as you cut off the waste wood. Support the piece so that the front face is underneath, and cut away the waste until you get close to the lines you have drawn. When you get close, it is necessary to use callipers (Fig. 4.4) to check the actual thickness at each station. Measure how many more millimetres need to be removed, write it in pencil on the workpiece at each station, and resume shaving the piece down until the thickness is correct to within 0.5 mm.

(Note: If you do not have callipers, it is easy to make a good pair [Fig. 4.4] using two pieces of sheet aluminium, or even plywood, bolted together.)

At the root, be sure to leave an area untouched (just as you did with the front face) for sandwiching between the hub disks.

Step 5. Smoothing out the section

You should now have a tapered, twisted blade, of the correct thickness. The cross section is just a crude parallelogram shape (shown bold in Fig. 4.5), which is not very aerodynamic. The final stage of carving your blade is to give it a streamlined airfoil 'section'.

Start by feathering off the trailing edge. Plane off wood from the

—A Cross Section of the Blade—

Fig. 4.5

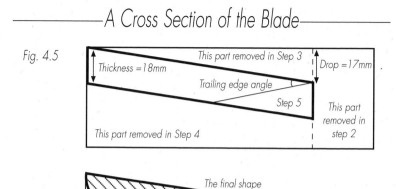

back (not the windward face) until you have a sharp edge, less than a millimetre wide, bevelled at the 20° angle shown. Set the work up with light shining onto the trailing edge, so you can easily see how wide it is. The finished edge should be under 1 mm wide.

Note: It may be helpful to make an angle template (Fig. 4.4), which you can fit over the trailing edge, to check that you have got it right, and adjust it accordingly. Again, this can be made from plywood or aluminium sheet material.

Finally, the section needs rounding off into a smooth 'wing shape'. Take care not to reduce the overall thickness. The thickest part should be at about 35% of the width back from the leading edge. Draw a line along the back of the blade at this thickest point, and avoid cutting the line.

Round the back of the blade off by continually removing the corners, running your fingers over the surface of the back of the blade or watching the way the light casts shadows as it rakes across the wood. Use sandpaper if you must, but a really sharp spokeshave, set very fine, is lovely to use.

Step 6. Assembling the rotor blades

Ensure that the thickness is the same all over each blade root. Reduce the large ones if necessary. The exact thickness does not matter as long as they are all the same.

—Assembling the Hub—

Fig. 4.6

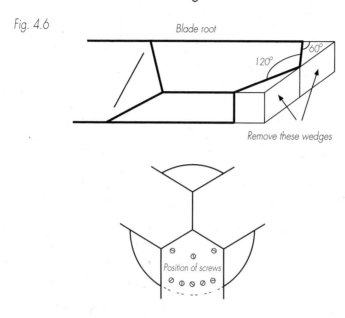

Blade root

120°

60°

Remove these wedges

Position of screws

Each blade root must be cut to a point (Fig. 4.6) to fit snugly at the hub. Measure the exact centre of each blade root, and draw lines out to the edges, at an angle of 60° to each edge. Mark them front and back, then cut along the lines.

The blades can now be laid out with all three roots fitting together. They will be supported in this position by the two plywood disks, one on each side. Make a pencil mark on each blade, 152 mm from the root (front and back), to help you to centre the plywood disks.

Drill and countersink holes in each disk for the screws (Fig. 4.6). I suggest 8 screws on each side of each blade. They must not obstruct the holes which you will need to drill, for bolting the rotor to the windmill. Check that the blades are equally spaced. Measuring the distance from tip to tip and adjusting them until equal is the easiest way to ensure a 120° angle between the blades. Check also that the tips are all the same height above the bench on which the plywood sits. This will ensure that they 'track' properly

The Finished Rotor

A 2.3m diameter rotor as described in the instructions.

(follow each other through space). If the blades do not track within 5 mm, there will be some 'dynamic' imbalance.

Drilling the bolt holes is best done with a drill press if possible. In any case take care to drill the holes square to the rotor.

While dismantling the hub for painting, take care to mark each blade for re-assembly. Use a drill to make a number of shallow dimples in each blade (one, two or none), and mark the disks to match.

More hints on wooden blade construction

The above procedure does not cover all shapes and sizes of blade. Here are some more hints:

Two bladed rotors can be built from a single piece of wood (Fig. 4.7). This saves work in constructing a hub. It is simple and strong.

The central portion of the piece of timber can be left full-size. Bolt the rotor to the generator pulley, then drill a large hole through the centre of the rotor, to accommodate a socket spanner when fitting the pulley to the shaft. The shaft nut must be locked with thread sealant to prevent it from working loose.

Two Bladed Rotors

Some two bladed rotors carved from wood.

Two Bladed Rotor

Fig. 4.7

Close-up view:

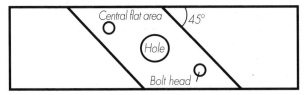

Central flat area   45°

Hole

Bolt head

Fins at the root

A glance at the blade designs in chapter four shows that it is desirable to have a wide section and a coarse setting angle at the root of the blade. If you cut such a blade from a single piece of

Leading Edge Protection

Fig. 4.8

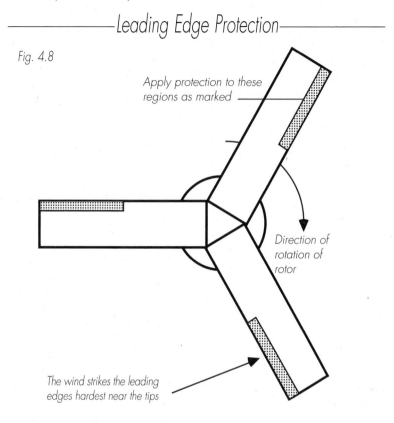

Apply protection to these regions as marked

Direction of rotation of rotor

The wind strikes the leading edges hardest near the tips

timber, it would be very wide and thick, and there would be a lot of waste. A simple solution is to use offcuts from the outer parts of the blade, glued on as supplements to the inner part. Build up the trailing edge with a fin to increase the chord width. Build up the windward face with another fin to increase the setting angle.

Painting and balancing the blades

High tip speeds cause rapid erosion of the blade material. The leading edges of the blades need special treatment (Fig. 4.8), either with epoxy resin or 'leading edge tape'. Leading edge tape is sold for the propellers of microlight aircraft (see the access details in the back for suppliers). It is easy to apply and to replace, and it offers very effective protection for a period of time. Apply the tape after painting.

If you are using epoxy resin, you should first plane off about 3 mm from the leading edges and rebuild them with a paste, mixed from epoxy resin and aluminium powder (or a similar filler). Polyester resin pastes such as 'Plastic Padding' are not so durable as epoxy. Apply the resin before painting.

The ideal protective surface would be a resilient (rubbery) finish. Good adhesive strength is also required. Silicone rubber can make a good on-site repair, but it is hard to produce a smooth finish.

Painting

Prime the wood carefully, and apply plenty of coats of gloss paint. Sand it well before the final coat. Household gloss paint may seem crude, but it has advantages over epoxy paints and varnishes.

Epoxy paint is completely watertight, which is not always an advantage. Water within an epoxy coating cannot escape, whereas other paints will breathe. If the blade is damaged, for example by fastenings biting into the wood near the root, then water will enter, and be centrifuged out to the tip, where it will collect, and swell the wood, until it bursts. Varnish degrades much faster than paint in ultra-violet light. (Varnished wood does look lovely though, for a while.)

Balancing

It is essential to balance the blades carefully. The aim is to ensure that the centre of gravity of the assembled rotor is exactly at the centre of rotation, i.e. the centre of the shaft. This is known as 'static balancing'. Dynamic balancing is not necessary, provided that you ensure that the tips of the blades 'track' each other. Rotor blades are axially thin, so static balancing is quite sufficient.

Balancing should be done indoors, in a large open space, free of draughts.

Pieces of lead flashing (from the scrapyard) make ideal balance weights. If very heavy weights are required, they can be shaped (from steel or lead), and tucked into the recesses between the three blades.

Here is one method of checking the static balance. The blade is poised on a sharp spike (Fig. 4.9), perhaps made from a 100 mm

─────────A Jig for Balancing the Rotor─────────

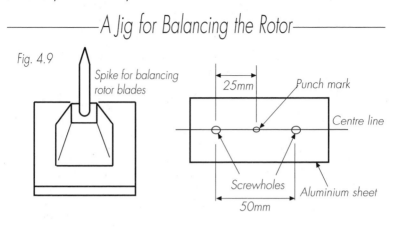

Fig. 4.9

Spike for balancing rotor blades

25mm

Punch mark

Centre line

Screwholes

Aluminium sheet

50mm

nail or similar, driven into a wooden support, and sharpened with a grinder. The spike engages with a punch mark at the exact centre of the rotor.

Make a jig from a small piece of aluminium sheet (Fig. 4.9) with a punch mark at the centre, between two screw holes. Position the holes precisely, at (say) a 25 mm radius from the mark. Make two corresponding holes in the back of the rotor hub, on an exact line through the centre of rotation. Screw on the jig, with the punch mark at the exact centre of the rotor.

Engage the spike with the punch mark and sit the whole thing on its stand. It will be unstable. Set the rotor level, using a spirit level. Lift off the spirit level gently, and observe in which direction the rotor falls. Add weights to the opposite side, until the rotor is capable of balancing momentarily on the spike, with no preferred direction to fall in. You need to place the spirit level both 'north-south' and 'east-west'.

Making the rotor is perhaps the most satisfying part of building a windmill, and it is a feasible task for anyone with simple tools, patience and enthusiasm. Finding a suitable generator to drive with your rotor may not be so easy though, as we shall see in the next chapter.

Chapter Five
Generators

The hardest part of designing a small windmill for electricity production is to find a suitable generator. In this chapter we look at some of the options.

What to look for

You need a reliable, low speed generator with good efficiency especially in light winds. Permanent magnet alternators qualify on all these counts, and they are by far the most popular choice in successful small windmill designs. They are discussed on pages 84 and 86. Brushless d.c. motors are just like these and an almost perfect choice too. They are discussed on page 85. However, there are other options.

An off-the-shelf machine is the ideal choice, for the obvious reason that you do not have to build it yourself. If it is mass-produced the production cost will be low, and also there will be surplus units available as scrap. The more common it is, the more easily you will find spares. Sadly, generators are seldom usable on windmills 'as found'. Part of this chapter is devoted to explaining how to adapt them. One exception is the 'dynahub' alternator for bicycles, which is perfect (if you can find one) for building tiny windmills.

There are other options beyond either re-using something 'as is', or building from scratch. In this chapter we shall look at some of the possible modifications which can be made to an existing

Flux around a Bar Magnet

Fig. 5.1

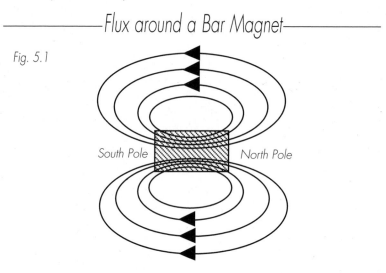

South Pole   North Pole

generator for lower speed. Often you can use a motor as a generator, and this widens the field quite significantly. Some d.c. servo-motors are suitable 'as found'.

Even if you build your own generator, you can choose some existing piece of mass-produced engineering as a starting point. For example, I have found vehicle wheel hubs and brake parts very useful for mounting permanent magnets and coils when building a windmill. More on this later.

How generators work
Magnets

A magnet has two poles, north and south. Flux emerges from the north pole and finds its way back to the south pole (Fig. 5.1). This is the 'magnetic circuit'. Flux loves iron, steel and other such magnetic materials. Not only does the flux attract them, but the amount of flux in the magnetic circuit will be much greater if the circuit is made of such materials.

Coils

Generators also contain coils of copper wire, often also referred to as 'windings'. The copper wire is coated with a thin enamel, which insulates each turn of wire from its neighbours. A coil is usually wound on a wooden 'former', and then taken off, built into

Stator and Rotor Configurations

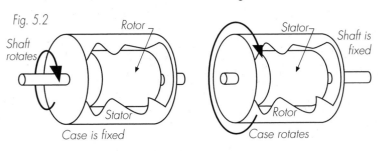

Fig. 5.2

Shaft rotates

Rotor

Stator

Case is fixed

Stator

Shaft is fixed

Rotor

Case rotates

---

the machine, and set in resin to form a solid lump.

A coil can be one of two types:

• Main, or output coils, in which the power is generated, or
• Field, or excitation coils, which need to be fed with current, in order to create a magnetic field in the machine. This is known as 'exciting' the machine.

Permanent magnet alternators need no field coils, as they are permanently excited.

Stator and Rotor

Generators have two parts: a stationary part called the 'stator' and a moving part called the 'rotor'. The stator is usually the outer part, encasing the machine, while the rotor is mounted on a rotating shaft in the middle (Fig. 5.2).

The opposite arrangement is possible though, and in fact it is quite common in windmills. The shaft is fixed, and the case rotates on it. This arrangement is known as 'case driven' as opposed to the conventional 'shaft driven' arrangement. If the windmill needs no gearing, the rotor blades can be bolted straight onto the magnet rotor.

Generators work by moving magnets past coils, or coils past magnets; it does not matter which moves. What matters is the relative motion. The magnet(s) can therefore be on the rotor or on the stator. They can also be on the inside or the outside.

The advantage of having the coils on the stator is that they are easy to connect to, without sliding contacts.

──────────────── *Flux Linkage* ────────────

Fig. 5.3

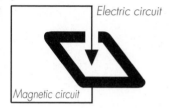

Electric circuit

Magnetic circuit

The flux around the magnetic circuit links with the wires in the electric circuit (the coil), like links of a chain.

Flux cuts wires

The magnets and coils in a generator are configured in such a way that the flux from the magnets passes through the coils (Fig. 5.3). In other words the magnetic circuit and the electric circuit link through each other.

In one position of the rotor, the magnetic flux through the coils is concentrated through the coil in one direction. As the rotor moves, the flux through the coil diminishes to zero and reverses completely. Flux ebbs and flows in an endless cycle, like the waves in the sea, cutting the wires as it goes. As the lines of flux cut through them, a voltage is produced in the wires. This is known as 'electromagnetic induction'.

Figure 5.4a shows a simple two-pole alternator. The shaft carries a magnet which rotates, dragging its flux through coils which are fitted into the stator core (see Fig. 5.4b).

Figure 5.5 overleaf is a graph of how the coil voltage changes over time, as the alternator rotor turns.

Maximising the flux in the machine

The magnetic circuit is often compared to the electrical circuit, in an analogy which also equates flux to current. The air gap in the magnetic circuit acts like the resistance in an electrical circuit. A large air gap limits the amount of flux around the circuit.

To keep the air gap small, the coils of our simple alternator are fitted into slots in the stator core. The steel between the slots provides a 'low resistance' path for flux passing through the coils.

A Simple Two Pole Alternator

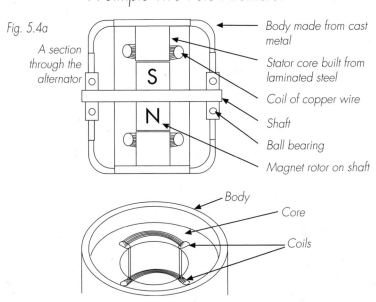

Fig. 5.4a

A section through the alternator

Body made from cast metal

Stator core built from laminated steel

Coil of copper wire

Shaft

Ball bearing

Magnet rotor on shaft

Body

Core

Coils

End View of a Two Pole Alternator

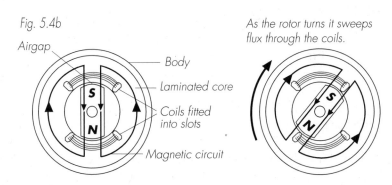

Fig. 5.4b

Airgap

Body

Laminated core

Coils fitted into slots

Magnetic circuit

As the rotor turns it sweeps flux through the coils.

Iron loss

The fact that the flux is changing in the core all the time affects not only the coils around it, but also the steel in the core itself. We do not want these 'side-effects' in the core; they waste power. They

Current from a Two-Pole Alternator

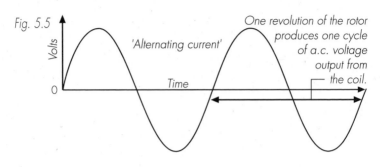

Fig. 5.5

Volts

0

'Alternating current'

Time

One revolution of the rotor produces one cycle of a.c. voltage output from the coil.

are called 'iron loss', and occur for two reasons:

- The iron is being magnetised and demagnetised at a rapid rate. This process involves hysteresis, and so takes energy. Special steels which are easily magnetised can be used to reduce hysteresis loss.
- The flux changes tend to produce circulating currents in the steel, following any conductive path which links around the changing lines of flux. A core built from flat laminations (insulated from one another) can be used to break up any such large circuit paths, minimising these 'eddy currents'.

Cogging

Slotted cores have teeth (for the flux) between the slots. As the rotor turns, the magnets are attracted to these teeth. The result is a pulsing torque effect, known as 'cogging'. This can cause difficulties with starting the alternator up, and can also cause noise when running. These effects can be minimised by skewing the slots slightly.

Multi-pole machines

So far, we have looked at a rotor with just two poles: north and south. A magnet has two poles, but there can be many more poles in a generator. The number of poles is always even, because there can never be a north pole without a south pole, or vice-versa. Figure 5.6 shows a four pole alternator.

Frequency

If there are two poles (Fig. 5.5), the voltage undergoes one complete cycle for every revolution of the machine. If there are four poles, you will get two cycles per revolution. The rate at which the voltage alternates like this is known as the 'frequency' of the supply. Frequency also varies in proportion to rotational speed (see windpower equations). If the voltage undergoes 50 cycles per second (as mains electricity does) then the frequency is '50 Hertz' (50 Hz).

Phase

Most alternators have more than one output coil. In 'single phase alternators' you can connect all the coils together to supply power to the same circuit. This is possible because the cycles in output voltage from the coils are all 'in step' with each other. Technically, we say that the coils are 'in phase' with each other.

Consider Figure 5.6 overleaf, where there are four poles and four coils. As the rotor turns, each coil faces a pole at the same moment. You can connect all these coils together in series (Fig. 5.7 overleaf), just as you would connect the individual cells together in a battery, to produce a higher voltage supply. Alternatively you can connect the coils in parallel to give more current (at lower voltage) than in series.

In generators where one coil is facing a south pole, and another is facing a north pole at the same moment, the connections to one coil simply need to be reversed, and the coils will work together.

If you are supplying a number of circuits, or if the supply is to be converted to d.c. for charging batteries, then it is preferable to use a 'three phase' alternator, with three sets of coils, all producing a.c. with the same voltage and frequency, but out of step. The coils are distributed on the stator (or rotor) in such a way that the poles pass one coil after another, in a smooth succession. Most electricians will associate the words 'three phase' with a 415 volt supply, but other voltages are perfectly possible. Car alternators have three phase windings, for example.

Figure 5.8 shows the stator of the four pole machine 'unrolled' into a flat rectangle, so that the layout of single phase coils is easier to follow. The position of the four poles (on the rotor, facing the

End View of a Four Pole Alternator

Fig. 5.6

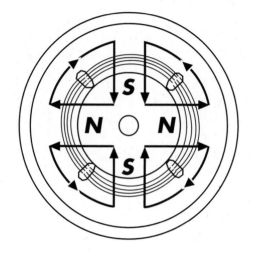

Coils in Series

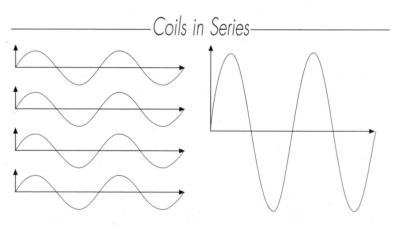

Fig. 5.7 Four coils in series give four times the voltage output.

coils) is shown 'N, S, N, S', to highlight the fact that the coils are all in phase.

For contrast, Figure 5.9 shows the layout of coils in a three phase machine.

There are six coils, in three pairs, shown joined up with thin lines. The first pair are in phase with each other, and can be connected in series to supply one circuit. The second pair can again

A Four Pole Single Phase Stator

Fig. 5.8

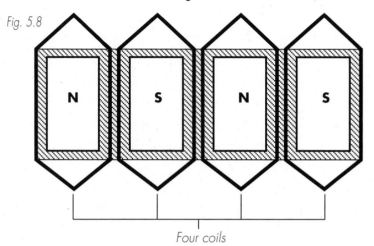

Four coils

A Three Phase Winding

Fig. 5.9

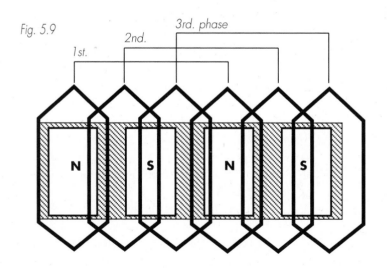

1st. 2nd. 3rd. phase

Votrage from a three phase alternator

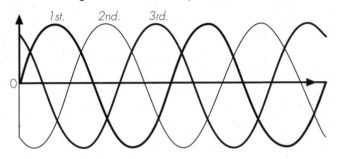

Fig. 5.10 How the voltage from each of the three phases varies with time.

be connected to each other, but their timing is slightly different from that of the first phase, so they must supply a second circuit. Again, the third pair will produce a third independently timed electricity supply from the machine.

Looking at the timing of the peaks and troughs of the three supplies (Fig. 5.10), we find that they all give the same voltage at the same frequency, but they are out of step with each other (out of phase).

Why bother? You may well ask. We now have a total of six wires ('tails') coming out of one machine. It's a headache to understand. There are good reasons, though, such as:

- In the single phase machine, the coils all want to be in the same place, and there are large areas unused. A three phase winding makes better use of the available space. This can result in a more efficient windmill.
- A single phase machine produces its power in pulses (Fig. 5.5), whereas the three phase machine is producing power continuously (Fig. 5.10). There is much less vibration. This can mean a quieter windmill!
- Three phase a.c. makes better use of cables than three separate single phase supplies (which would need twice the amount of copper wire to do the same job). This is of less importance for battery charging systems, because d.c. has even lower cable losses than three phase a.c.

────── Ways to Wire Up Windings ──────

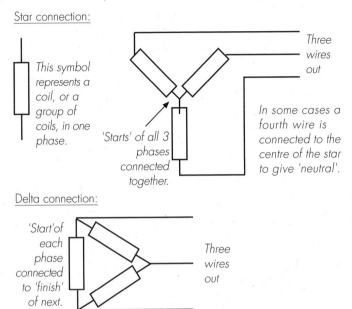

Star connection:

This symbol represents a coil, or a group of coils, in one phase.

'Starts' of all 3 phases connected together.

Three wires out

In some cases a fourth wire is connected to the centre of the star to give 'neutral'.

Delta connection:

'Start' of each phase connected to 'finish' of next.

Three wires out

Fig. 5.11 The two ways to wire up three phase windings.

Star and Delta connections

It is normal to wire the three circuits together. We can't connect them in series (or parallel) because of the time lag, but they can share some wires in common. There are two options for connecting the tails (Fig. 5.11).

Voltage

Voltage is the 'push' which drives current around a circuit. The components of an electrical system must all be designed to work at the same voltage (rated voltage). We are accustomed to thinking in terms of electricity supplies having a particular voltage, and we take it for granted that this will remain constant. You may not realise what an effort goes into maintaining the constancy of the mains supply voltage.

The voltage is determined by the rate at which the wires cut flux. Voltage can be increased by increasing any of the following:

• Speed of rotation;

- Strength of flux;
- Number of turns per coil.

But what do we mean by the voltage? Voltage is not a simple commodity with a.c. supplies, because it is changing all the time, dropping to zero, reversing, etc. A voltmeter will give you a nice steady reading, but what does it represent? (See Fig. 5.12)

In fact, the meter will tell you the 'rms' voltage. 'Rms' is short for 'root mean square', which is the value of d.c. voltage which would make a bulb glow with the same brightness. If you are planning to charge a battery from this a.c. supply, then the peak voltage (about 40% higher) may also be important.

Internal volt-drop

So far we have assumed that the output of the machine is 'open circuit', so that no current is drawn from the machine. The open circuit voltage is also known as the 'electromotive force' (emf) of the generator. Once you start to draw power from the generator, there will be 'volt drop' due to losses in the windings.

Consider a '12 volt' dynamo (see later for a description). As the speed of the dynamo increases, so does the voltage, until cut-in speed is reached at, for example, 1,200 rpm. This is where the open circuit voltage is 12 volts. At 2,000 rpm it may actually produce 20 volts open circuit. But when you connect it to a 12 volt battery, it only produces 12 volts at the same speed, while pushing (say) 20 amps of current through the battery. 8 volts are used up inside the dynamo in overcoming internal resistance. Figure 5.13 is a graph of voltage and current versus speed, to show what is happening.

In real life, other factors would further confuse the situation, for example changes in flux and in battery voltage.

D.c. supplies

Most small windmills which produce electricity are used for charging batteries. If the windmill uses an alternator, the a.c. must be converted to d.c. This is done with semiconductor devices called 'diodes', which act as one-way valves for electric current. A number of diodes are assembled into a 'bridge' circuit as shown in Figure 5.14. A three phase bridge can be built on the same principle (see later).

Which Voltage is it?

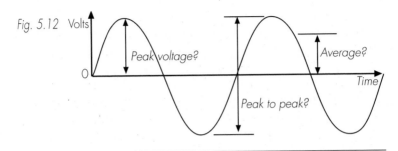

Fig. 5.12

Voltage is Modified with Current

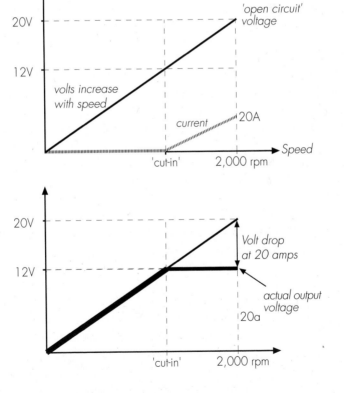

Fig. 5.13 How the open circuit voltage is modified when the current flows.

—Bridge Rectifier Circuit Diagram—

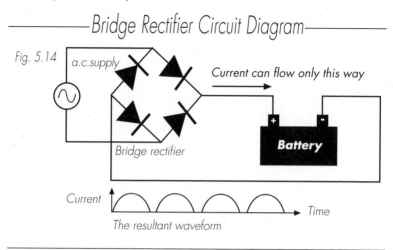

Fig. 5.14

a.c.supply

Current can flow only this way

Bridge rectifier

Battery

Current

Time

The resultant waveform

As current passes through each diode there is a volt-drop of about 0.7 volts. To charge the battery, current has to pass through two diodes in the bridge. So in order to charge a 12 volt battery, about 13.4 volts input is required. Thus, more than 10% of the power is lost in the diodes. Bridge rectifiers must be mounted on a 'heat sink' (a piece of aluminium with fins) which will carry away the waste heat and prevent burn-out.

Current starts to flow in pulses as soon as peak voltage exceeds the battery voltage and diode volt drops. As the input voltage rises, the current grows stronger and more steady, especially in the case of three phase supplies.

Brushes and sliprings

Earlier we looked at a permanent magnet alternator with all the coils on the stator (Fig. 5.4). Most generators and motors also have coils on the rotor. To connect these coils to an external electrical circuit, such rotors are normally fitted with sliprings. A slipring is a smooth copper surface, built into the rotor, and turned on a lathe. Sliding contacts called 'brushes' are pressed against the revolving surface of the sliprings by springs. Originally a brush-like arrangement using fine wire braids, brushes are nowadays made from slabs of carbon.

For example, the field coil of a car alternator is on the rotor (Fig. 5.15). Field current is fed to the coil via the brushes and sliprings.

The Rotor of a Car Alternator

Fig. 5.15

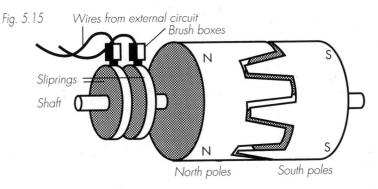

Wires from external circuit
Brush boxes
Sliprings
Shaft
N
N
S
S
North poles
South poles

Cross section of rotor

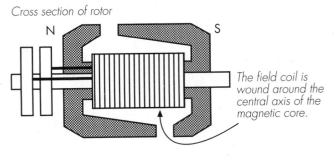

N
S

The field coil is wound around the central axis of the magnetic core.

This current also passes through a regulator circuit, which prevents the output voltage from exceeding the safe maximum for battery charging, irrespective of speed. If permanent magnets are used, there is no need for brushes, but there is no opportunity to control the output in this way.

Brushes wear out, over a period, and the contact surface of sliprings can become pitted by sparking (especially if overloaded with current, or contaminated) so a generator with no brushes will be more reliable.

Commutators

In the old days, before semiconductor rectifiers came along, it was very difficult to charge batteries from a.c. You had to use a special d.c. generator, called a dynamo.

The body of the dynamo (the stator) carries the field magnets (Fig. 5.16). Inside is a special rotor, called an 'armature', which has

Sectional View of a Dynamo

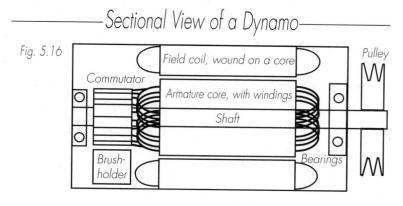

Fig. 5.16

Commutator

Field coil, wound on a core

Pulley

Armature core, with windings

Shaft

Brush-holder

Bearings

---

many coils and a 'commutator'. The trick for producing d.c. is in the commutator and brushes which connect the armature coils to the external battery circuit.

The name commutator literally means 'switcher' and that's exactly what it does. The series of coils on the armature is wired to segments of copper on the commutator. The brushes are carefully positioned relative to the field poles such that they always connect to the coils which are cutting flux in only one direction. The result is that d.c. comes out of the machine.

Commutators are also used in some motors, called 'd.c. motors' or 'universal' motors.

Changing the speed of generators

Generator speed is the biggest issue in windmill design. Most generators are designed to run too fast to be successfully coupled directly to the windmill rotor, but there are some tricks you can use to reduce the operating speed.

We have seen that the voltage produced by a coil depends on the speed, the flux and the number of turns. There is a maximum (saturation) limit to the amount of flux. So to operate them at lower speed, we must either work at lower voltage, change the coil grouping from parallel to series (or delta to star) or use more turns per coil.

Sadly, all of these options reduce the rated (maximum) power output in watts, without reducing the copper loss. For example, you can operate a 24 volt alternator at half speed, charging a 12 volt battery. But if the current is the same, you will only get half the power, because power is voltage times current. Copper loss (due to the effect of current flow in the main windings) will remain the same. The copper loss is now twice as large a fraction of output power, so the efficiency has suffered.

If we want to keep the same efficiency while reducing speed, then we need to cut the current as well as the voltage. The rated power is then reduced by the square of the reduction in rated speed. For instance, at half the speed we get one quarter as powerful a machine. This is the price of low speed operation; you need a larger generator in relation to the power output. It does not really matter how heavy the generator is once you have got it up in the air, but its efficiency is crucial, so in most cases the extra weight is worth it.

In the above example we changed the voltage to achieve reduced speed of operation. We can alter the number of turns per coil or the grouping of the coils (into series or into star) to restore the original voltage, but this leads to higher internal resistance, and the current must be cut down by the same factor, giving no increase in power output. The reason for making such alterations is to maintain compatibility with a certain battery voltage, while working at lower speed (and lower power).

If there are also field coils, the constant loss in them will become more significant as the power rating is reduced, so it will not be worth modifying the generator to lower speed operation unless it has good efficiency to begin with.

Types of generator

Mains voltage generators are not covered in this book as they are seldom used in small windmills.

Car alternators and dynamos

Car alternators are a very popular choice for newcomers to windpower. They are readily available, they have the right sort of power rating, and they are designed for charging batteries. Nevertheless they are not ideal when building windmills. The part-load efficiency is very poor, because of the need for current in the field coil. Modern car alternators are built for cheapness, light weight and high rpm.

Dynamos also need field current but they can be a good choice for a windmill, if you can find an old enough one. This may seem a strange thing to say, but older dynamos were designed for lower speed of operation, with the emphasis on efficiency rather than on maximum power. For a direct drive windmill you will need a big, heavy dynamo (over 20 kg for 300 watts). We are talking about pretty old-fashioned equipment, and they are getting hard to find.

Some important facts about dynamos and alternators

Car alternators and dynamos are self-excited, which means that above a certain speed they cut in and excite their own fields. In both cases, the field coil(s) are connected in parallel with the output. Cut-in happens spontaneously. The magnets hold a weak remanent magnetism, which produces a small voltage. This produces a small current which enhances the existing field and produces more current, and so the voltage builds up. Full voltage is achieved in about one second. Alternators do not cut in so easily. The field coils often need a small current, provided through the 'indicator lamp' (see later).

Above the cut-in speed, dynamo output current increases steeply with increasing speed. Maximum output is achieved at a realistic speed for windpower, whereas with alternators you need to reach three to four times the cut-in speed before rated output can be obtained (Fig. 5.17).

Car alternators actually limit their own current (by inductive reactance in the windings), which means that (if adequately cooled) they are almost impossible to overload, whereas a dynamo will burn out if pushed too hard. Alternator rotors are also capable of surviving very high speeds, which would destroy the armature of a dynamo.

Old dynamos have excellent bearings, so you can comfortably mount a fairly large (2-3m diameter) rotor directly onto the pulley. Car alternator bearings are less impressive.

There is no essential difference between a dynamo and an electric motor. Current will flow from the battery to the dynamo and turn the windmill during calms unless a 'blocking diode' is fitted to prevent reverse flow. An alternator cannot

Power/Speed Characteristics

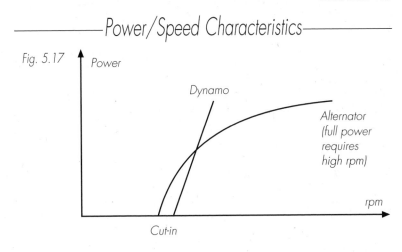

Fig. 5.17

Power

Dynamo

Alternator
(full power
requires
high rpm)

rpm

Cut-in

---

readily be motored, and reverse current is blocked by the rectifier.

Dynamos do not mind which terminal is positive or which is negative (the polarity). The dynamo 'learns' its polarity when you connect it to a battery, by motoring it or simply 'flashing' the field circuit. Alternators are badly damaged if you connect them up with the wrong polarity.

Alternators do not mind in which direction you rotate them (unless their cooling fans do). A dynamo will only work in one direction. It will motor the same way. You can reverse the direction of rotation by reversing the polarity of the field coils (swapping the wires).

Dynamos are almost silent in operation, whereas most alternators will make a humming noise.

Working with dynamos

Dynamos need regular maintenance, at intervals of about one year, perhaps longer if well treated. This involves removing the armature from the body, cleaning out the carbon dust, cleaning the surface of the commutator (taking great care to keep it concentric; a lathe is sometimes used to skim it), cleaning and greasing the bearings as required. The brushes must be able to move freely in their boxes, and the commutator surface must be completely free of grease. It's a dirty, fiddly job, but very necessary. If the commutator is not functioning properly, the dynamo will fail to self-excite,

Modifying a Dynamo's Wiring

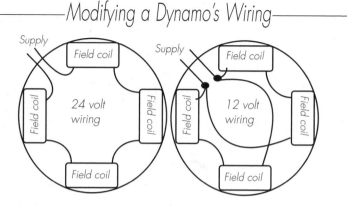

Fig. 5.18 Modifying a dynamo's field wiring for lower voltage operation (end view of the open dynamo body).

and the windmill will consequently overspeed.

You can almost halve the rated speed of a 24 volt dynamo by using it to charge a 12 volt battery. The field coils need to be rewired in parallel rather than series to work properly at the lower voltage. Figure 5.18 shows a typical layout.

If you are lucky enough to find a low speed dynamo, take care not to overload it. High currents will lead to sparking at the brushes and accelerated wear. In the worst case it will burn out the armature, which is tricky to rewind.

Car alternators

If you want good results, it is necessary to understand the wiring of the alternator (Fig. 5.19).

The current from the stator windings passes through a three phase bridge rectifier, which converts it to d.c. for charging the battery. There is one large terminal on the alternator, which is connected to battery positive. The negative is connected to the case (earth). Never connect the battery 'back to front' or you will destroy the rectifier! Never run the alternator fast without a battery connected, or the voltage will rise to dangerous levels, which may also damage the rectifier.

A car alternator requires a supply of current to the field coil, via the brushes. This could be drawn from the battery, but it would continue to drain the battery when the engine (or windmill) was

Car Alternator Circuit Diagram

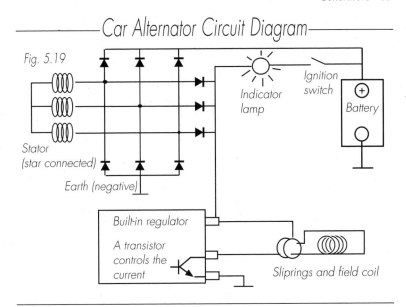

Fig. 5.19

Stator (star connected)

Earth (negative)

Indicator lamp

Ignition switch

Battery

Built-in regulator

A transistor controls the current

Sliprings and field coil

stopped. Instead, we have a special rectifier, with nine diodes, which provides an independent positive supply from the stator windings to the field coil.

This extra terminal on the rectifier is usually marked 'IND' for connection of the indicator lamp. The lamp will only glow when there is no voltage available from the alternator. When generating, there is the same voltage at each end of the bulb, and so it will extinguish.

Most car alternators have an internal regulator, which controls the output voltage by means of the field current. This is not suitable for use on a windmill and should be bypassed and discarded. It cannot easily sense the battery voltage, and also it would unload the windmill, causing runaway. Regulate the battery voltage with a shunt regulator instead (see chapter seven).

Modifications to car alternators

Rewind the stator coils with more turns, to work at the same voltage but at lower speed. Study the existing coils as you remove them. Wind replacements using enamelled winding wire of about 70% of the original diameter and as many turns as you can fit, which should be more than double. Now the alternator will cut in at an acceptable speed for direct drive by a 1.5 metre diameter rotor, but the power output and the efficiency will be reduced.

It is also possible to modify the rotor by fitting permanent magnets. The efficiency and simplicity will be enhanced in this way, but power output will be lower still.

Permanent magnet alternators

Permanent magnet alternators have similar characteristics to car alternators (above) except that they have no field coils. They are the best choice of generator for small scale windpower because they are simple and efficient. The magnets on the rotor require no brushes or sliprings, so the only parts which can wear out are the bearings. There is no requirement for current to excite the field, so losses are at a minimum in light winds.

The problem is that they are hard to find. Here are some of the few possible sources (and their drawbacks):

- Exciters for brushless synchronous generators (expensive);
- Bicycle and motor cycle alternators (low power rating). The bicycle ones are useful for very small windmills (5 watts), as they have quite a low operating speed, but motorcycle alternators are designed for high rpm, and are not easily adaptable for direct drive;
- 'Mighty Midget' brand of welders (tend to become demagnetised);
- Chinese manufacturers of small wind turbines. Try the Internet. Cheap but tricky to deal with.

Motors used as generators

As we mentioned above, motors and generators are very similar, and are often interchangeable with little or no modification. In fact, motors always generate a voltage when they are turning. This voltage, known as 'back emf', is less than the supply voltage, so the net effect is that a current still flows from the supply to the motor. As the motor speeds up, the back emf rises, the supply current falls, and so the motor is regulated to a speed dictated by the supply voltage.

The difference between motoring and generating is just a matter of speed. Say you connect a dynamo to a 12 volt battery. There will be a cut-in speed at which that particular machine generates exactly 12 volts. Force it to rotate faster, and it will generate more emf, causing current to flow into the battery. Slow it down, and it will produce less, allowing current to flow from the battery into what is now a motor.

The same principle applies with a.c. motors. In fact, most large wind turbines connected to the grid use induction motors (see more later) as generators. If there is only a light wind, then the machine will tend to wobble to and fro, between motoring and generating, quite smoothly and naturally. The control system of the windmill connects or disconnects it from the grid, at a windspeed chosen for maximum net benefit.

Permanent magnet motors

Windscreen wiper and fan motors in cars are mainly of the commutator type, with permanent magnets. The speed is generally much too high for practical use in windpower, but toy windmills could be built with them for very low voltage applications. Starter motors sometimes have permanent magnets these days, but the brushes are only designed for brief periods of operation. Older starter motors have series field windings which are not suitable.

There are larger permanent magnet d.c. motors available, with lower speeds and better efficiency. Many of these have commutators, and behave much like dynamos (above). If such a motor is available at low cost (scrap washing machine motor perhaps) then you can well use it on a windmill. Try to avoid high currents which may burn it out or damage the commutator.

Brushless d.c. motors

There are low speed, brushless, permanent magnet servo-motors available nowadays, which are just about perfect for use as generators on small windmills. Brushless motors are just like permanent magnet alternators. In order to work as motors they are fed with a.c. exactly in phase with the rotation. This is produced by special inverters. The low weight is achieved by the use of expensive 'rare earth' magnets, which create a very high flux density in the motor.

Applications include machine tools, robots, military and

medical equipment. As these things are always being scrapped, there must be cheap, brushless d.c. motors to be had for those who know where to look. Even some washing machines are being built with brushless d.c. motors these days!

Induction motors

Also known as 'asynchronous motors', these are both the commonest kind of motor to be found and the hardest kind to understand. They are cheap to produce, brushless, and low on maintenance. An induction motor has all its windings on the stator. In fact the stator is similar to that of an alternator, with a core built from a stack of laminations. The rotor is a simple cylinder of laminated steel containing an embedded 'squirrel cage' of aluminium bars, with an aluminium ring at each end. Field current is induced in these bars by the currents in the stator windings, but a full explanation is beyond the scope of this book. Suffice it to say that although there are no brushes or field coils, a supply of field current is still required, in the form of an out-of-phase 'reactive' current to the stator.

When induction motors are used as stand-alone generators, they can be self excited using capacitors. This is rather a black art. There is a very good book on the subject called *Motors as Generators for Micro-Hydro Power*, IT Publications (available from CAT bookshop). Windmills are a more difficult case than hydro, because power and speed vary so widely, but it can be done. The fact that geared motors are available at low cost in a range of sizes simplifies the problem of building such a windmill. 'Energy efficient' motors are well worth the extra expense.

Building a permanent magnet alternator from scrap

If the cost of permanent magnet alternators is prohibitive, then building your own is a very practical option. Here are some hints.

Ceramic magnets

Permanent magnet materials have advanced in recent years. There are many kinds to choose from: ceramic, alloy, rare earth... all greatly superior to the old steel magnets. The cheapest and easiest to use are the ceramic magnets, made from ferrite material.

Axial versus Radial Field

Radial field   Axial field

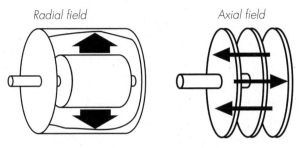

Fig. 5.20 The arrows show the direction of the magnetic flux.

Other magnets have much higher flux density and lower weight, but ferrites are relatively cheap and stable. Look for material called Feroba 3, or Ferroxdure 330, depending on the supplier.

Suppliers (see Access Details at the back of the book) can offer catalogues with standard sizes of magnet blocks. They can also cut the blocks to suit your requirements, and supply them to you pre-magnetised. They use a diamond saw to cut them. You can also grind them with a stone-cutting disc.

To make an alternator, you will need to find or build a suitable machine, with a rotor and stator. Glue the magnets to the rotor in such a way that a magnetic circuit passes through coils which you have wound and fitted in the stator (see Fig. 5.3). The magnetic circuit must be made of suitable materials. Steel or cast iron will do for the rotor, but the stator core (if any) must be laminated. Aluminium and plastic are not suitable for the magnetic circuit.

There is no limit to the number of different shapes of machine you could dream up. They can be classified into radial or axial field machines according to the direction of the flux in the air gap (Fig. 5.20).

Radial field machines

Most electrical machines have radial field, similar to that in Figure 5.4. The car alternator and the induction motor are further examples. You can convert them to permanent magnet alternators by fitting magnets to the rotors, using a large number of rectangular magnet blocks to create an approximation of a circle. Detailed plans are available (see Access Details).

Laminations for Alternators

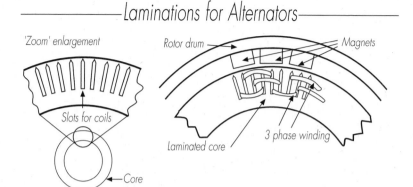

Fig. 5.21 The best shape of laminations for 'case driven' alternators.

Some windmill manufacturers in Holland use induction motor bodies as the basis for their alternators. They use many poles and wind the stator with many more coils than an induction motor. The main difficulty with this arrangement is that the magnet blocks are subject to high centrifugal forces, combined with frequent, large changes in temperature. These conditions are quite demanding for the glue.

The other radial field arrangement is to put the rotor on the outside. Two big advantages of this 'case driven' arrangement are that the centrifugal forces press the magnets against the rotor, rather than pulling them off; and that you can use the brake drum or even the wheel of a vehicle as a rotor. Car wheel bearings are cheap, and substantial enough to support rotor blades bolted to the hub.

To build the stator for such a machine, we need to find a laminated core, onto which we can fit the output coils. The outside of this core must come close to the magnet faces, leaving only a small air gap for the flux to jump across. A wide air gap means lower flux density, but where the flux comes from ferrite magnets there is only a small loss of performance if the gap is about one millimetre. The advantage of a wide gap is that there is less danger of the rotor and stator making contact due to slack bearings or distortion, as often happens in the real world.

Figure 5.21 shows the best kind of laminations to use for such a machine, but they are hard to find these days. Most motors

—An Alternative Approach—

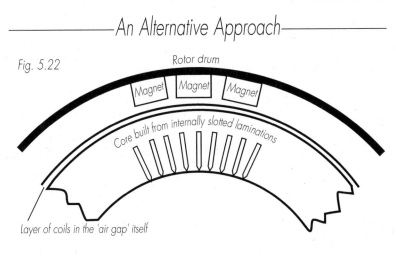

Fig. 5.22

Rotor drum

Magnet Magnet Magnet

Core built from internally slotted laminations

Layer of coils in the 'air gap' itself

---

obtainable as scrap are induction motors, with slots on the inside of the core. However, there are specialist suppliers of laminations, or complete stators, for DIY permanent magnet alternators of this sort (see Access Details).

Figure 5.22 shows an alternative approach, using the laminations from a common induction motor. For example, a brake drum with a 254 mm (10 inches) bore from a van with twin-wheel rear axle (e.g. Ford Transit) can be used with magnet blocks 20 mm thick glued around the inside, as shown above, and 203 mm (8 inch) laminations from an electric motor. The resultant air gap (about 5 mm) is large enough that you can glue coils to the surface of the stack of laminations, 'in the air gap' as it were. There are detailed plans for doing this, available from the author (see Access Details).

This arrangement is less powerful than an alternator with externally slotted laminations would be, because there is less space available for windings and because more of the flux from the magnets 'leaks' from one pole to the next without cutting any wires, but the arrangement is very efficient for low power outputs (in light winds). The iron losses are less than with the externally slotted laminations, there is no cogging, and starting is easier.

An 'Air Gap' Alternator

Fig. 5.23

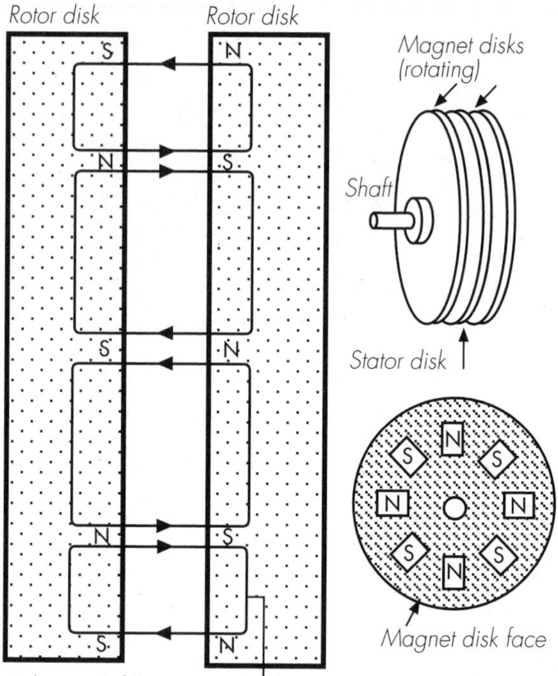

Rotor disk   Rotor disk

Magnet disks
(rotating)

Shaft

Stator disk

Magnet disk face

*Side view of the rotating
disks showing the magnetic
circuit crossing the air gap
twice each way.*

Axial field machines

'Axial field' means that the flux lines crossing the air gap are parallel to the axis of the shaft, jumping from one disk to another (Fig. 5.20, page 87). The commonest type of axial field permanent magnet alternator is the 'air gap' alternator, as used (for example) in 'Rutland' windmills.

Figure 5.23 depicts the magnet disks of an air gap alternator, seen from the edge. There are two disks, with several poles (8 or more) each, facing each other. The north pole on one disk faces the south pole on the other (and vice versa). Flux passes across the gap in both directions. Large loudspeaker magnet rings are often used, magnetised with many poles using a special jig. Or the self-builder

Coil Connections

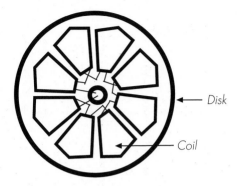

Fig. 5.24 Series connection of coils in a single phase stator disk of the 'air gap' type. Every second coil has its connections reversed.

can glue pre-magnetised ceramic blocks to the faces of steel disks. One advantage of the latter arrangement is that the magnet blocks act like the blades of a fan, helping air to circulate past the stator and cool it.

The stator disk sits in the gap between the two magnet disks. It consists of a set of coils embedded in resin (a polyester resin, as used for fibreglass construction, will serve). In the single phase version, there is one coil on the stator for each pole on a magnet disk. The coils can be connected in series to produce higher voltage, but every second coil needs its connections reversed, because the flux direction alternates (Fig. 5.24).

It also is possible to build a three phase version. There are three coils for every four poles (e.g. six coils for eight poles). Every third coil is connected in series to make a phase group (no need to reverse any this time). Join the starts of all three phase groups together, and use the three finishes as output to the rectifier. (See also Fig. 5.11, 'star grouping'.)

Even with the 'air gap alternator' there is a choice between mounting the rotor on the shaft or on the case (Fig. 5.25). Axial field machines are usually 'case driven', but this is not the best arrangement, because the stator disk is only supported at the centre, where it is easily warped or cracked. The shaft-drive configuration is more robust.

Fig. 5.25 'Air Gap' Alternators

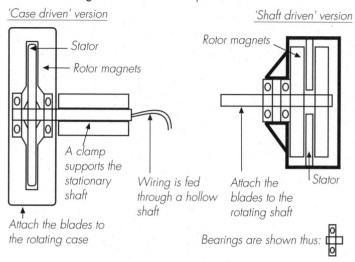

Fig. 5.25 Drive arrangements for 'air gap' alternators. Sectional views of the two types.

The 'air gap' alternator has no laminations in it, so:
• There are no iron losses;
• There is no 'cogging' and minimal starting torque required;
• You can build one without needing laminations.

However the thin stator disk is difficult to cool, so it can be burned out easily.

Design hints

There are plenty of 'windmill plans' available (see Access Details), but for those who want to design their own alternator, using available bits or personal brainwaves, here are some hints.

Air gap diameter

The best shape for a low speed alternator is 'fat'. The circle described by the air gap should be large. By increasing the diameter, you increase the speed at which magnets pass coils. This improves the efficiency for a given weight of magnets and coils, at a given rpm.

The limit on making the alternator 'fat' comes when the rotor cannot be made sufficiently rigid to prevent contact with the stator. Magnetic forces, gyro forces, and slack bearings all contribute to this risk. The operating frequency also limits the diameter.

The number of poles

If the air gap diameter is large, then there is room for plenty of poles. Your choice of the number of poles is fairly arbitrary, because a few wide poles will push very little more flux through the coils than many narrow ones will. For each alternator there will be an optimum, depending on the following factors:

- Frequency depends on the number of poles. Higher frequency (more poles) is helpful if you want to use transformers, but also leads to higher iron loss;
- Copper loss depends on the width of the coils and is less if the poles are smaller;
- Some flux fails to pass through the stator but jumps to the next-door pole. The amount of flux which leaks in this way is more where there are many poles and a wide air gap;
- The magnetic circuit needs to be more heavily built in order to carry more flux without saturating, if the pole faces are large.

The shape of the coil

The coil should be shaped so that as each pole passes the coil, most of the flux will link through it. The thickness of the coil will depend on the space available. For example, the coils in an 'air gap alternator' will be the same thickness as the stator disk. Coils in a laminated core must be designed to fit the slots, with due allowance for insulating liners. If the coils overlap (as in most three phase windings) you must provide enough slack for this.

The number of turns

This needs to be chosen so as to give you the desired voltage at the desired rpm. The voltage/speed performance is not always easy to predict, especially where the design has not been used before.

An Axial Field Alternator

Magnet rotor and a single test coil for an axial field alternator. (Built by Eric Archbold).

A Brake Drum Alternator

Bits of a DIY brake drum alternator.

See the Equations for a 'rule of thumb' formula for the number of coil turns which produce 12 volt cut-in at a given rpm. Divide these turns among the coils in parallel grouping.

If the cut-in speed turns out to be too low, or the voltage is too high, you can reconnect the coils in delta or parallel. But it is better to use a replacement winding with fewer turns. Delta or in parallel windings may suffer from parasitic or 'harmonic' currents which circulate between one coil and another.

Thickness of wire

Use the thickest wire you can comfortably fit into the available space, so as to minimise the copper loss. If the thick wire is too stiff you can use two or more thinner wires bundled together.

Unlike cables, winding wires do not have a specified current which they can carry without overheating. This is because the cooling depends entirely on the geometry of the winding. Where the stator has a large surface area and good ventilation, it will be able to dissipate more heat before reaching the temperature where the insulation is damaged.

Use the formula for predicting copper loss in wires (see Appendix) to estimate what fraction of the power is lost in the windings. Check that the efficiency is acceptable (above 50%, say). You can also check how much current the stator can handle, by passing this current through it in a bench test, and measuring the temperature rise. On a windmill, the cooling will often be better.

In conclusion

The generator is the heart of the wind turbine, and the hardest part to get right. Given a good generator, you are more than half way there. You can make blades according to the details in previous chapters. You will also need good controls, as described in the next two chapters, to get the best from your generator.

Chapter Six
Mechanical Controls

In addition to its main task of converting windpower to electricity, the windmill has to adapt to circumstances. It must face itself into the wind. It also needs to protect itself from the violence of winds greater than the rated windspeed. This must happen automatically, because nowadays it is quite unrealistic to expect anyone (even yourself) to be able to look after it full time. Sooner or later, there will be an occasion when a gale gets up while you are out.

Larger wind turbines have computer-driven control systems, which operate servo-motors, hydraulic motors, and all sorts of paraphernalia. On a plant of that size it is economic to provide this and to keep it all maintained. The same approach could work on a small windmill, but the cost would be too high. Reliability problems would be the worst part of the cost.

Small wind turbines need simple, passive controls where possible. Always assume that any moving parts will seize up or wear out. If anything can vibrate, then it will do so until it falls off. Lightning may strike at any time, and on some sites this is quite a likely occurrence. If water can get into a space then it surely will, and often it gets in where it strictly should be impossible. Simplicity is all-important. Perfection is not the state where nothing further can be added, but rather where nothing further can be taken away.

Good controls can greatly increase the power production of a windmill, by helping it to reach its full capacity, and by keeping it there in all weather. Shut-down mechanisms are sometimes easier to devise, but systems which keep the windmill in full production through adverse conditions are far better.

Facing into the wind

Most windmills of under 10 kW capacity use a tail to keep them facing into the wind. Vertical axis machines have no concept of facing the wind (having no 'face'). A few HAWTs run downwind, which means that a tail is neither necessary nor convenient. But small, upwind HAWTs use tails.

Tail design

Every HAWT has a 'yaw bearing', on which it sits, giving it freedom to rotate. The vertical line through the centre of the yaw bearing is known as the 'yaw axis'. The tail is a vane on the end of a boom. The vane catches the wind, pulls on the boom, and swivels the machine about this yaw axis to face the wind.

The amount of turning moment (see Glossary for definition of moment) which the tail needs to produce will depend on such things as:
- Friction in the yaw bearing;
- Aerodynamic forces on the rotor (thrust, self-orienting forces...);
- The tendency of the centre of gravity to swing downhill, if the tower is not quite vertical.

The yaw moment produced by the tail is simply the side force on the vane multiplied by the length of the boom. A longer boom will compensate for a smaller vane. The side force on the tail vane will depend on the area of the vane, and on the windspeed (squared). For details see the Equations.

As a rule of thumb, the actual boom should be about equal to the length of one blade, i.e. half of the rotor diameter, with a vane centred on its end. A cross-piece or other sort of stiffener may be useful, depending on the rigidity of the vane.

The area of the vane will depend on the job it has to do. If the yaw bearing is very free-moving, the rotor set centrally on the windmill and the tower vertical, then the vane can be quite small. Tail vanes are rarely smaller than 3% ($1/30$ th) the swept area of

Shaping the Tail Vane

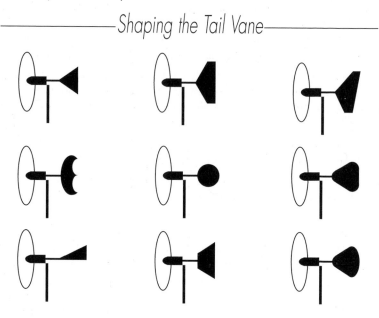

Fig. 6.1 Some ideas for shaping the tail vane.

the rotor. For example, a 2 metre diameter rotor will have a swept area of about 3 square metres, so a tail area of about 0.1 square metres would be the minimum. This would be a vane 300 mm square or similar. (Note that you would need a larger vane for a furling windmill; see later.)

A tall thin vane is slightly more effective than a long, low one, but the shape makes very little difference to function. The 'look' is important though. If you are a mechanical rather than an artistic person, then it may be wise to get help with this. See Figure 6.1 for some ideas.

Cut the tail vane from any sheet material, but beware of thin or brittle materials such as aluminium sheet. Plywood makes a very durable vane.

Tails are quite surprisingly liable to fall off because they pick up on vibrations produced by the windmill and resonate with them. Never underestimate the effects of this perpetual waggling. Gust loads, combined with rotor wobbles and generator hum, add up to a wide spectrum of surplus mechanical power acting on the tail. The tail and its fastenings need to be substantial, for safety.

Yaw drive mechanisms

Larger windmills use other methods to face the wind. On the old corn-grinding mills, yawing was known as 'luffing' and at first it was done manually, like everything else. The windmill had a tail-pole projecting down to ground level, from the back. The miller used a simple winch to manoeuvre the tail pole to the best position. Even nowadays, a length of scaffolding tube clamped on the back of a prototype windmill is an excellent short-term control system. If the rotor runs out of control you can yaw it away from the wind by hand. Hard hats are advisable for this job, in case bits drop off in the process!

With the development of cheaper gearing, the 'fan tail' was introduced to corn-grinding mills in the mid 18th century. A fan rotor is mounted on the back of the main windmill, at right angles. The fan is connected by a series of gears and chains to a gear which yaws the main windmill back to the square-on position if the wind comes from one side. The fan tail is a highly geared device, so it works slowly and surely, which is perfect.

Fan tails have largely been superseded by electrical or hydraulic drives on modern windmills, but their simplicity is a big advantage. For example, they are still popular on stand-alone systems, where there is no other source of power to drive a yaw motor before the windmill starts up.

Avoiding overload

It is neither economic nor wise to fit the windmill with a big enough generator to convert into electricity all the power from the highest windspeed it will ever see. But if it does not absorb all the power, then the rotor will overspeed, resulting in excessively high centrifugal forces, unacceptable noise and vibration — living dangerously, in other words. If the rotor diameter exceeds about one meter, you will need to prevent this overspeed.

One approach is to use centrifugally operated 'air brakes', but this is rather like motoring with one foot on the brake and the other on the throttle: not very good for the car in the long run. A better approach is to avoid capturing the power in the first place. Keep a low profile and ride out the storm.

There are two common types of 'governing systems' which

Pitch Control

Dunlite centrifugal pitch control mechanism.

aerodynamically limit the power captured by the rotor:

- One which yaws or tilts the whole machine, so that the rotor becomes skew to the wind. In the side-facing position it literally presents a 'low profile'. Windspeed through the rotor is reduced and so is the power captured.
- One which adjusts the pitch of the individual blades. Twisting the blades changes the angle of attack, reducing the lift.

Governing systems which move the whole rotor are simpler to construct but slower in operation than pitch controls, so they do not work so precisely. The power output is much less smooth on turbulent sites where the wind constantly varies in speed and direction. Pitch control is fast acting and can give a nice smooth output, but it is very difficult to set up. It requires a hub with moving parts which must also be strong enough to withstand all the punishing forces and moments to which the blade roots are subjected, while still moving reliably and correctly to adjust the pitch. Springs are included, too. Pitch control is not covered in detail by this book.

Turning away from the wind

There are a host of proprietary systems, with clever names like 'autofurl', 'furlmatic' or 'hinged vane safety system'. They all work on very similar principles, activated by wind pressure on the rotor itself. As the windspeed increases, the thrust on the rotor increases too, until it reaches a point where it activates the furling mechanism, and the windmill turns away from the wind.

The windmill either yaws away from the wind sideways, or tilts back so that it faces upwards. In either case the rotor becomes skew to the wind. This effectively reduces the component of windspeed through the rotor, limiting the speed and the power output. In practice, the power output depends on the component of thrust force along the rotor axis, regardless of the angle of the rotor to the wind. So by keeping this force constant, we can govern the windmill.

The secrets of 'side-facing' furling systems

Let us look at an example where the windmill yaws sideways. We use the thrust of the wind on the rotor to drive the furling movement. The thrust on the rotor is centred on its axis. If this axis is offset from the yaw axis, then the thrust creates a yawing moment, turning the windmill away from the wind (see Fig. 6.2).

In normal windspeeds we do not want the rotor to yaw sideways; we want it to face the wind squarely and catch all the power. So we build a large enough tail to withstand the yawing moment caused by the offset, using a vane area around 10% of the rotor swept area. The tail vane is offset by an angle of attack of about 20° to the opposite side from the rotor axis offset. This tail produces a restoring moment which equals the yawing moment from the rotor thrust.

As the wind increases, there comes a point when the windmill delivers its maximum output (the 'rated windspeed'). Beyond this point the tail must automatically swing aside, and allow the windmill to yaw away from the wind. In other words, the tail's 'restoring moment' must have a strictly defined upper limit, beyond which it yields and allows the rotor to yaw.

Yawing the rotor off the wind reduces the windspeed through the rotor, reducing the axial thrust, and an equilibrium is reached,

Yaw Governing System

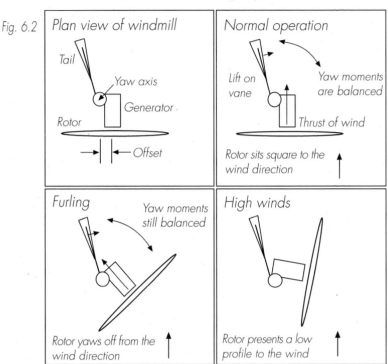

Fig. 6.2

Plan view of windmill

Tail
Yaw axis
Generator
Rotor
→| |← Offset

Normal operation

Lift on vane
Yaw moments are balanced
Thrust of wind
Rotor sits square to the wind direction

Furling

Yaw moments still balanced
Rotor yaws off from the wind direction

High winds

Rotor presents a low profile to the wind

giving reduced power. You do not necessarily want the windmill to stop working altogether. A constant restoring moment (pulling the rotor back into the wind), throughout the range of movement, will ensure a constant thrust on the rotor, giving a constant output independent of windspeed.

In many cases the restoring moment becomes weaker as the windmill furls. Such windmills furl too easily in high winds, (Fig. 6.3) and the power output drops off. This is safer than overspeeding, but it is also unstable. On a turbulent site, it may yaw the windmill suddenly, putting high gyroscopic bending stresses on the blade roots.

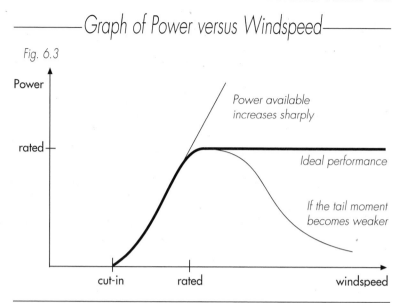

Graph of Power versus Windspeed

Fig. 6.3

Building furling tails

The larger the rotor offset, the larger the tail must be to hold the machine straight in normal winds. But the minimum offset you can use is 4% of the rotor diameter (as a rule of thumb). For example a 2 metre diameter rotor should have an offset of at least 80 mm. (Note: If the offset is too small, the rotor may start to get ideas of its own. At particular tip speed ratios it will seek the wind. This 'self orienting moment' can defeat the governing system entirely. A rotor with a very small offset will yaw itself into the wind without the help of a tail under some conditions.)

The obvious design is a spring-loaded tail, which folds up when the force on the spring is exceeded. But a spring is vulnerable to weathering and fatigue. Nor does it lend itself to producing a constant restoring moment. The tension in the spring increases as the spring stretches. The radius of action of the spring (its distance from the hinge) will also change.

The best way to make a reliable, self-furling tail is to use gravity instead of a spring to pull the tail into its normal working position. This is achieved by mounting the tail on an inclined hinge. The tail falls down against a stop under its own weight, in the normal (low windspeed) position (Fig. 6.4). The restoring moment does vary

Gravity Operated Furling Tail

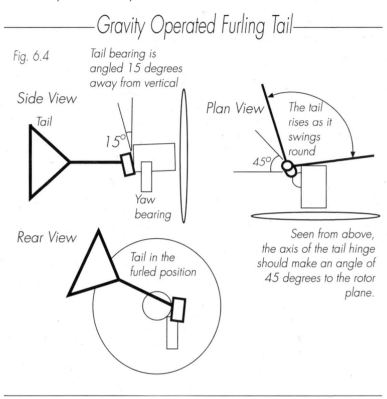

Fig. 6.4

Tail bearing is angled 15 degrees away from vertical

Side View

Tail

15°

Yaw bearing

Plan View

The tail rises as it swings round

45°

Seen from above, the axis of the tail hinge should make an angle of 45 degrees to the rotor plane.

Rear View

Tail in the furled position

---

somewhat, reaching a peak at the mid point of the tail's swing, but in practice this variation is usually tolerable.

A simpler approach is to use a fixed tail boom, with a vane suspended under it on hinges (Fig. 6.5). This tail is less precise in operation, and more highly stressed in storms because it projects across the wind, and is subjected to its full fury, with a vane flapping off it.

Tilt-back systems

The other arrangement for skewing the rotor to the wind is to mount the windmill on a bearing which allows it to tilt-backwards and face the sky (see Fig. 6.6 overleaf). Tilt-back systems of governing are as popular as yaw governing with furling tails. The choice usually seems to depend on the designer's previous experience. Tilt-back systems can exhibit strange gyroscopic movements under turbulent conditions.

Hinged Vane System

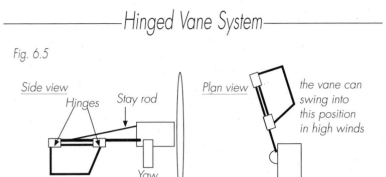

Fig. 6.5

Side view — Hinges — Stay rod — Yaw bearing

Plan view — the vane can swing into this position in high winds

The tilt-back system again needs some sort of restoring moment to hold the rotor to the wind. Often the weight of the generator and rotor hold it in place at first, but, as it rises, the weight goes over the pivot point. The restoring moment diminishes, and the windmill tends to tip right back and slow right down. This position is unstable; it will crash forward again, and this will happen repeatedly on turbulent sites.

A spring can be used to pull it forward, compensating for the loss of restoring moment, but springs are not desirable on windmills. If you must use a spring, use a stainless steel one which is very over-sized and under-stressed. Pay special attention to avoiding fatigue of the fastenings. Figure 6.7 overleaf shows two other ways to produce the desired extra restoring moment. Both have their drawbacks.

Shut-down systems

Shut-down systems are comforting to have, but their use is an admission of defeat, because a shut down windmill is a missed opportunity. It is bad enough that windmills will not work in calm weather; they ought to work full time in windy weather!

Good governing systems should cope automatically with any windspeed. But if there is a serious loss of balance in the rotor, or an electrical failure, then an emergency stop can be useful. Also, when raising and lowering a windmill on its tower in a breeze, the rotor must be stopped.

——————Tilt-Back Governing System——

Fig. 6.6   _Side View_

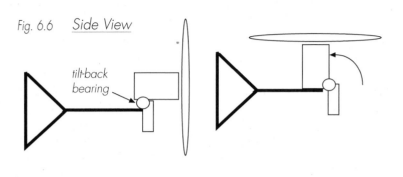

tilt-back
bearing

——————Tilt-Back Governors without Springs——

Fig. 6.7 Method 1:
using a counterweight

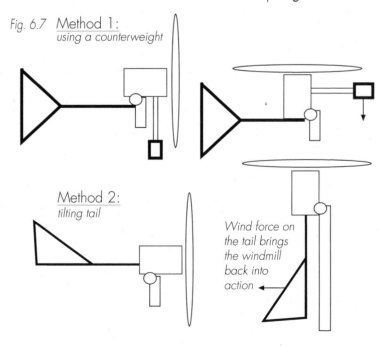

Method 2:
tilting tail

Wind force on
the tail brings
the windmill
back into
action

Mechanical brakes are rare in small windmills. A brake which is large enough to stop the mill in a good breeze needs to be well built. Half measures are not of much use. A good brake:

- is expensive;
- takes up space in the 'nacelle' or clutters the shape of the windmill;
- requires maintenance and testing if it is to be relied upon, and so it is generally best avoided.

Brake switches and heaters

Permanent magnet alternators keep working right down to very low speeds because the flux is always present. If you short circuit the output, by connecting all a.c. wires together (using a 'brake switch'), the alternator will pump large currents around its windings and the cables. Even at 'no volts' the copper loss will waste enough power to stall the rotor blades very effectively.

The brake switch is a good way to prevent unwanted start-up, but not always a good way to actually stop a windmill. There are two potential problems.

Firstly, some alternators, once they are in full power production, will generate very little more torque with the brake switch on than off. They limit their own current (by internal inductive reactance). Windmills with this sort of alternator cannot be stopped unless the speed drops below a certain rpm. (Hint: switching on the right size of heater on the a.c. wires will ensure that it stalls during the next lull in the wind.)

Secondly, other alternators deliver such high currents when short circuited that the shock puts an unacceptable load on the rotor blades, causing cracks at the roots.

Brake switches cannot be relied on in the case of an electrical failure, such as when the cables become disconnected from a windmill. It will run away, because the load has been removed. The same fault is likely to have disconnected the brake switch too, so that will not help.

Despite these drawbacks, the brake switch is a very useful parking brake, and since it costs nothing (except a decent, big switch), it is well worth having, if only for use during erection of the tower. Alternatively, a heater, which is tripped on in the event of overspeed, makes a good automatic shut-down.

Furling cables

Rather than using a mechanical brake, you can use the existing furling mechanism to shut the windmill down through a furling-rope, which furls the tail, or tilts the windmill back to face the sky.

Here are the drawbacks:

- The furled windmill cannot be relied on not to start up during erection of the tower, when it is lying on its side;
- Brake cables are very prone to corrosion and will probably seize up unless regularly checked and lubricated;
- The furling cable follows the same route to the ground as the power cable. The two are very likely to become entwined. One simple solution is to use the electrical cable from the windmill as a furling rope. Just pull the cable, but not too hard — you don't want it to break!

You can also use a servo-motor to furl the tail, or to apply a brake of some sort. A geared windscreen-wiper motor driving a big screwed rod can move a nut with slow determination. It can be driven by a circuit which detects overspeed, or even a full battery. Windmills have been controlled in this way, but there is just too much to go wrong.

Chapter Seven
Electrical Controls

Electrical controls are another important link in the chain between wind energy and useful power. As with mechanical controls, the key to good reliability is to keep things as simple as possible.

Electrical and electronic equipment needs to be mounted on a panel or in a box. If you leave it hanging off the battery terminals or lying in long grass under a windmill, then you are inviting corrosion. Support all cables as they approach terminals or connections, to prevent movement. Mains voltage connections should definitely be housed in boxes to prevent contact with unwary fingers.

Electronic controls serve two distinct purposes (often using similar circuitry for each):

- Controlling the load on the windmill to optimise the speed, and maximise the energy capture by the rotor blades;
- Controlling the current through the battery to protect it from damage and keep the voltage within safe limits.

Load control: the key to good performance

The rotor will operate best at a particular tip speed ratio. During start-up, the tip speed ratio will be below this optimum point, so the loading of the generator must be very light at low speed. This allows the rotor to accelerate through stall until it reaches its best speed.

Battery charging systems

Fortunately, a battery charging circuit loads a generator quite nicely without any controls. Hardly any power is required to turn the generator slowly, because no current flows in the output circuit until generator voltage exceeds battery voltage. By the time this 'cut-in' speed has been reached, the rotor is running at a high enough tip speed ratio to deliver the necessary power.

Heating systems

Where power is fed directly to electrical loads such as heaters, some sort of electronic controller is necessary. If you simply connect the generator directly to a heater, you will have a nice simple electrical system, but the heater will start to draw power as soon as a voltage appears on the circuit, stalling the rotor.

Essentially, what the controller must do is allow the windmill to start and gather speed, until it reaches its design tip speed ratio. Only then is the heater connected. The windmill will immediately slow down under load, so the controller must then disconnect the heater again to prevent stall. In fact the heater may need to be switched on and off quite rapidly to maintain optimum loading on the windmill rotor.

The crudest sort of controller (which we could call a 'crash bang' controller) will simply switch on the full heating load, using a relay or 'contactor'. This would be a cheap solution, but very rough in operation.

There are several better solutions. For a start, it is better to use electronic rather than electro-mechanical switching to control the power to the heaters. Semiconductor switches have no moving parts, and they can be very reliable and quite cheap.

It is much better to load the windmill gradually rather than 'all or nothing'. There are two ways to do this. Either 'phase' the heater(s) in gradually, or switch them on one at a time. In either case, the controller senses the speed of the generator by monitoring the voltage (or even the frequency), and applies more load to the windmill as it speeds up. This loads the rotor blades optimally.

Phase Control

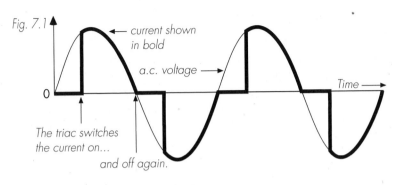

Fig. 7.1

current shown in bold

a.c. voltage →

Time →

0

The triac switches the current on...

and off again.

Pulse Width Modulation

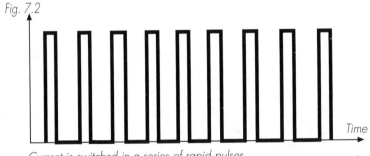

Fig. 7.2

Time

Current is switched in a series of rapid pulses.
Wider pulses carry more power to the heater.

Phase control

Electronic switching devices called triacs are often used to switch a.c. Once triacs are switched on they will not turn off again until the flow of current stops or reverses (see Fig. 7.1). The amount of current in the circuit can be controlled by delaying the start of each pulse, using triacs. (This is how 'dimmer switches' work.)

Pulse width modulation (PWM)

A more popular approach to heater control is to rectify the current into d.c. and switch it into very rapid pulses with a

—Multiple Heater Control—

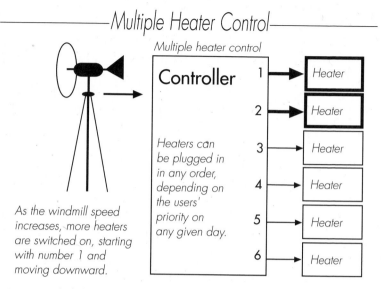

Multiple heater control

Controller

Heaters can be plugged in in any order, depending on the users' priority on any given day.

1 → Heater
2 → Heater
3 → Heater
4 → Heater
5 → Heater
6 → Heater

As the windmill speed increases, more heaters are switched on, starting with number 1 and moving downward.

Fig. 7.3 Multiple loads under stepped control.

transistor (Fig. 7.2). MOSFET transistors and IGBTs are good at this sort of d.c. switching, but they are less robust than triacs, and they do 'blow up' from time to time. Designing PWM controllers is a job for an electronics expert.

The pulses are switched at a high frequency (say 5,000 Hz) which sounds to the human ear like a high pitched whistle. The amount of power going to the heaters is controlled by regulating the length of the pulses.

Stepped load control

The above methods of control (sometimes known as 'proportional' control) give nice smooth results, but they can be tricky to design, and they can cause radio interference. An easier approach for the amateur electronics enthusiast is to use a number of small loads (heaters) and switch them in one after the other (Fig. 7.3). Similar to the 'crash bang' approach in its simplicity, this approach can work well provided that there are enough heaters. Then the load comes on in small steps which do not impose too much shock on the blade roots.

It is well worth using triacs as switching devices on such a system. They are robust and cheap. Triacs do not necessarily have to come on in mid-cycle (as they do in phase control systems). In fact it is possible to control them via 'zero crossing' circuits which switch them on exactly as the voltage crosses zero. This prevents radio interference. A more convenient (if costly) switching device is the 'solid state relay' (SSR) which is a single unit incorporating a triac and its driver circuitry.

Wire the heaters to plugs inserted into sockets in the control panel. By juggling the order of the plugs in the sockets, you can decide which rooms to heat first. Lower priority heaters will only be heated if the wind becomes strong enough. Or use small controllers built into the heaters themselves. Set each to switch on at a different voltage (or frequency) threshold.

Types of heating load

Ideally, a heating windmill should produce 230 volts a.c. or similar, so that off-the-shelf heating elements can be used. Voltage from windmills is seldom very steady, and it will be better if it is below 230 volts, rather than overloading the heater(s). Mains voltage heaters can work at lower voltages quite safely, but power varies with the square of the voltage. This means that at half of the rated voltage you only get one quarter of the rated power.

Convector heaters and storage heaters work well on windpower. Radiant heaters are noisy when fed with high frequency pulsed d.c., as the heaters resemble loudspeakers in their construction. Fan heaters are unlikely to work well under conditions of pulsed operation, reduced voltage, or constant stopping and starting.

Where the windmill produces three-phase a.c. (as is usual), there will be the option of running a separately switched single-phase heater off each of the phases, but then there is more current on one phase of the alternator than another, and this leads to increased vibration and noise. This problem of balance can be avoided by rectifying the power into d.c. after switching it with triacs. Heater elements will work quite happily on d.c. rather than a.c.. There are two things to note though:

- d.c. is more dangerous than a.c. in terms of shock hazard;
- thermostats are designed for a.c., so they may burn out on d.c. (They can be wired to separate, sensing circuits.)

If you are working at non-standard voltage and are willing to pay a little extra, you can buy high power resistors from electronics catalogues (see Contact Details) and make your own heater. Farnell now stock 300 watt resistors. The aluminium-clad ones work well for making low wattage water heaters. Just strap them directly onto the hot water cylinder. Do not forget to bond the plumbing to earth and enclose any live conductors, though.

Direct a.c. power supplies

The wind is unreliable, so it is rarely used for direct supply of a.c. circuits, but it can be done. To ensure a reliable supply, a large windmill is needed compared to the likely user load. Speed can be controlled by mechanical or electrical controls. You can use centrifugal pitch control, but it is more common to use heaters connected through a controller as described above. The heat is a useful by-product.

To run a.c. appliances you will need to provide them with something like their rated voltage and frequency. The quality of control required depends on what you intend to use the power for. Lighting needs a steady voltage to work reliably and acceptably. To run lighting directly from a windmill, you will need to regulate the supply voltage quite precisely using an automatic voltage regulator with proportional rather than stepped control. Stepped control causes too much flicker.

Motorised appliances, such as power drills, saws, food processors, vacuum cleaners, etc., will function with reasonable success over a range of voltages and frequencies. You can drill holes and grind metal using mains voltage tools at 150 volts quite satisfactorily. In light winds this may bring the windmill to a halt, however.

Power conversion using transformers

The commonest way to alter the voltage of a supply is to use a transformer. Transformers work on a.c., not d.c. If used at below their rated frequency you may have problems unless the voltage is also reduced. This suits windmills which run at variable speed, where both will drop together.

If you have a long cable run between the windmill and the

battery, you will either need a very thick cable, or a high voltage. With a transformer, you can use a 230 volt windmill to charge a 12 volt battery. You may encounter problems with start-up of the rotor, unless there is automatic disconnection at low voltage, because the impedance of the transformer is low at low frequency, and so a current will always flow in the circuit.

The extra cost of the transformer may be as great as the cost of the cable, but using a high voltage windmill has other advantages:

- You can dump power into a mains voltage heater when the battery is full;
- You can run power tools, blenders etc. directly from the wind;
- You can adapt the performance of your generator to suit the windspeed. By switching to alternative transformer tappings, you can take more power at higher rpm without increasing copper loss.

What batteries like best

Batteries are expensive, and your storage capacity is limited by the size of the battery. When the limits of the storage capacity are reached, then decisions must be made, either manually or automatically, to divert power through new channels.

Batteries need a little consideration. Like anything else you depend on, they repay some tender loving care. What batteries like best is being charged steadily and gradually with whatever current holds their voltage at about 14 volts (for a 12 volt, lead-acid battery).

There are two basic rules of battery management:

- don't let the voltage get too low;
- don't let the voltage get too high.

You can assist the batteries by adapting your habits to suit the wind, doing energy intensive tasks on windy days, and practising careful energy conservation during calms. But this can be burdensome if taken to extremes, and you need automatic controls too, if only as a back-up.

In what follows we shall assume that we are talking about a 12 volt system, when quoting voltages. For instance, battery heaven is 14 volts. If the system voltage is 24 volts, then you should double the voltage figures given, so heaven is at 28 volts instead. The voltages quoted are only guidelines. They will work OK for most lead acid batteries, but battery voltage depends on load current and on temperature as well as on battery type and condition. If you have the gel-type batteries with solid electrolyte (sometimes known as 'dryfit') then it is more important to prevent the voltage from going too high. Check the specifications of your battery.

Preventing low voltage

The most common cause of damage to batteries on windpower systems is being run flat, and then left flat. Batteries hate this. Voltage can be used as a simple indication of the state of charge. Do not discharge a 12 V battery below 11.5 volts unless you can immediately recharge it.

To prevent over-discharge, you can:
- Simply keep an eye on the voltage and manage the system yourself;
- Use a low voltage alarm;
- Have a low voltage cut-out, which automatically disconnects the TV and other inessential loads (Do not kill all the lights, or crash the computer without warning).

The instant the cut-out operates, the voltage will jump up slightly to the open circuit level. To prevent the cut-out from chattering back in and out again it is a good idea to build in some 'hysteresis' between the cut-out and the cut-back-in point. This has nothing to do with hysterics! It simply means that the threshold is different going back up from going down. For example, you could set the cut-out to restore power when the voltage rises above 12.5 volts again. A manual reset button is also a good idea, to restore power after you have switched some unnecessary loads off.

Take care when wiring up the cut-out. Do not disconnect the batteries and leave the loads connected to the windmill! A windmill with no battery connected can put out dangerously high voltages, which are almost certain to damage your appliances.

Relay puzzle

There is an interesting dilemma over the cut-out relay. If the relay is energised when the voltage is low (using 'normally closed' contacts) then it will draw power from the battery just when you can least afford to do so. If the relay is energised when the voltage is OK, ('normally open') then it will be wasting power all the time. This is only a problem with very small windpower systems, as the relay only uses about 1 watt.

Charging voltage controllers

If the charge current (amps) is greater than one tenth of the amp-hour capacity of the battery, then it is essential to have some kind of automatic control.

Floating volts

As noted earlier, batteries mostly love to be 'floated' at 14 volts. When a battery is very low it will take a large charging current to bring it up to 14 volts, and there will probably not be sufficient wind to achieve it. As the battery approaches full charge however, it will only require a small current to reach and maintain this voltage.

Charging regulators in cars simply limit the charging current from the alternator to prevent the voltage from rising above 14 volts. On windpower systems it makes more sense to divert the power into a heater, rather than blocking it at source, which may cause overspeed of the rotor.

Equalising charge

If the voltage rises above 15 volts, the batteries may bubble violently and overheat, causing damage if prolonged. Never do this to a gel-type battery. However it is good practice to overcharge normal batteries like this (keeping an eye on them) if they have been deeply discharged. This is known as an 'equalising charge' because it ensures that all the cells of the battery reach full charge.

Consider the loads, too

High voltage shortens the life of appliances. Halogen bulbs are expensive and so are television sets, so it is well worth having an automatic control to prevent over-voltage on systems with valuable equipment connected.

General arrangements of shunt regulators

The electronics involved in battery voltage controllers is very similar to the electronics used for heater control, as described in a preceding section of this chapter. To prevent the battery voltage from rising higher, you need to dump power into a heater, known as a 'dump load' or 'ballast load'. This type of regulator is known as a shunt regulator, because it diverts the current into a parallel or 'shunt' circuit.

You can use relays or semiconductors to control the heaters in a shunt regulator. Again, the crash-bang approach is not ideal. A number of small heaters work better than one big one. The presence of the battery will delay the swings in voltage. You could switch a heater on at 14.5 volts and off again at 13.5 volts, to give a delay of some seconds in between. Controllers like this can be completely modular and independent of each other. You simply connect more of them as your system expands.

It is also possible to dump the power from upstream of the rectifier. This has some benefits:

- A.c. is easier to switch. You can, for example, use triacs (but you need two if the supply is 3-phase). Triacs are more robust than transistors;
- If the shunt regulator malfunctions there is no danger of draining the battery, as there would be if it were connected directly to the battery;
- The battery is not 'cycled' by the dump load. It can be argued that the dump load shortens the battery life by draining power from it;
- The windmill will slow right down (in lulls) and stall (which may or may not be an advantage, depending on whether the dumped power is of use).

Dumping at mains voltage through an inverter

If the system includes a powerful inverter, then you can run mains voltage heaters on that. It is easy to switch mains voltage heaters as you can use smaller relays/semiconductors.

This may cause the lights to flicker, though. Also, if the inverter fails or shuts down due to overload or if there is a thermostat on the heater, then your dump load will no longer be available, so you

Circuit for Two Windmills

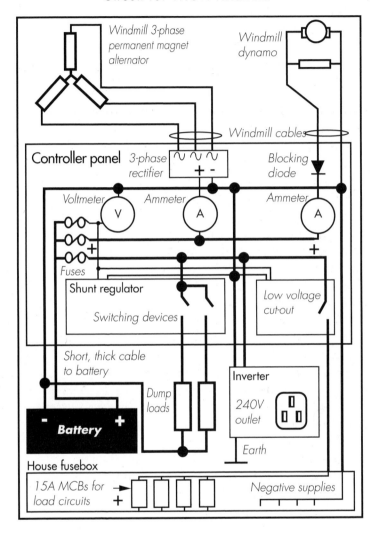

Fig 7.4 Overall systems circuit for a system with two windmills.

will need another dump load waiting in the wings to prevent the voltage from going too high. Nevertheless, it may be worth dumping into an immersion heater when you can, because this is so much more useful than the battery voltage heater.

Chapter Eight
Towers

I could write another book about towers alone, but here are some brief remarks.

Make the tower as high as you safely can, because the quantity and quality of the wind improves with height. A windmill hidden among trees will never achieve its full potential.

Types of tower: guyed and free-standing

Nobody likes guys. They are ugly, intrusive, and vulnerable to damage. Free-standing towers are greatly preferable, but unfortunately they are much more highly stressed, and so heavier and more expensive. In the end, guyed towers are the commonest choice, because they are cost-effective. So the rest of this chapter is only about guyed towers.

How strong is strong enough?

You must ensure that all guy ropes, shackles, anchors, etc., are big enough to take the strain. Here are the steps towards arriving at a suitable size.

Firstly, estimate the maximum forces which will occur in the life of the structure. Erection and storm force winds are usually the worst cases.

Then calculate how these forces will apply to the component which you are designing. To translate a force on a windmill into a tension in a guy rope, you need to use a technique known as

Forces on a Tilt-up Tower

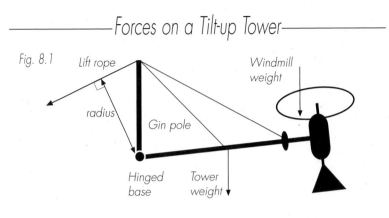

Fig. 8.1 Lift rope Windmill weight radius Gin pole Hinged base Tower weight

'taking moments about a point'. The obvious point to take moments about is the base of the tower. The top guy point is also worth looking at.

Lastly, apply a safety factor of five times between the 'safe working load' (SWL) and the breaking strain (see table 8.2 on page 125). A 'proof load' of twice the working load may be applied to check the strength of the arrangements if the breaking strain is not known exactly.

Erection

You can easily calculate or measure the weight of the windmill and of the tower itself. Erection is not always a smooth process, though. There may be jerks and bounces which double the actual forces. As this is a dynamic load, you should apply an extra safety factor of two.

In this book we shall look only at the 'tilt-up' tower. The base is hinged to allow the tower to swing up or down freely. The lifting rope passes over a 'gin pole' which ensures an adequate radius of action for the lifting force (Fig. 8.1).

The radius of action of the lift rope is roughly equal to the length of the gin pole. The longer your gin pole, the less force is required to lift the windmill and tower. The worst stage is where the tower is near horizontal (Fig. 8.1). As it rises, the radius of action of the weights diminishes.

Storm Forces on a Guyed Tower

Fig. 8.2

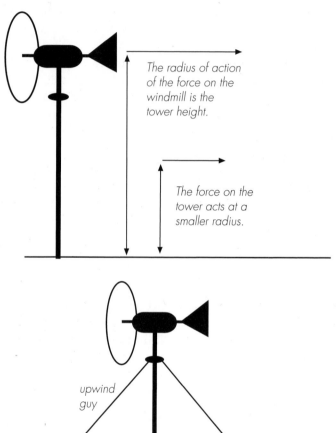

The radius of action of the force on the windmill is the tower height.

The force on the tower acts at a smaller radius.

upwind guy

radius of action

'guy radius'

Storm loading

The storm loading on the tower alone depends on the windspeed and the frontal (or 'projected') area which the tower presents to the wind. Assuming 50 m/s (110 mph) as your 'maximum gust', you can expect a force of 250 kg/square metre of exposed surface.

If the mill is shut down or turning slowly, then you can use the projected area to calculate the thrust on the windmill. If it runs but yaws away from the wind, the maximum thrust will be limited by the control system.

You can equate the overturning moment of the wind thrust on a guyed tower to the moment exerted by the tension in the upwind guy (Fig. 8.2).

The beauty of using a guyed tower is that all the bending moment is taken by the guys. There is no need to embed the base of a guyed tower in the ground. It will only need a foundation pad to prevent it sinking in.

Hands-on tower erection
Waterpipes

The simplest and commonest tower for a small windmill is a guyed tube. Steel tubing is easy to work with, being light to erect and easy to cut, drill and weld. Tubing also has the advantage that the cable from the windmill can be safely enclosed within it. Scrapyards are a good source of steel tubing in all sizes. Table 8.1 shows the sizes appropriate for different sizes of windmill.

Never rely on the threaded part of the tubing, as it can easily break under bending stresses.

You can make a simple but effective yaw bearing for a windmill by simply slipping a larger size of tube over the end of the tower tube. Figure 8.3 shows details of the welded construction of a yaw bearing and furling tail bearing similar to the one in Figure 6.2 (page 102) and the photo on page 58. Note that the cable simply hangs down the middle of the tube, and the top can be covered by a cut-off plastic washing-up liquid bottle. It's low-tech, but it works very well!

The alternator mount will depend on the alternator used, but an offset would be required to operate the furling system.

——Table 8.1 Tube Sizes and Rotor Diameters——

Windmill rotor diameter	Tube nominal bore suggested	Tube actual overall diameter
1 metre	'inch and a half'	48.3 mm
2 metres	'two inch'	60.3 mm
3 metres	'four inch'	114.3 mm
5 metres	'six inch'	165.1 mm

——————Fabricating from Steel Pipe——————

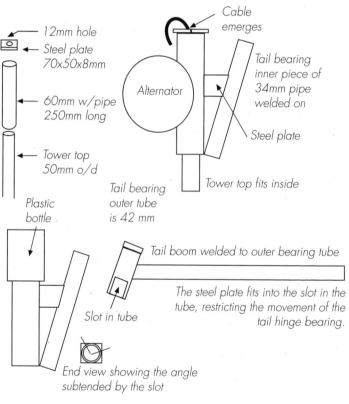

12mm hole

Steel plate
70x50x8mm

60mm w/pipe
250mm long

Tower top
50mm o/d

Plastic
bottle

Cable
emerges

Tail bearing
inner piece of
34mm pipe
welded on

Alternator

Steel plate

Tower top fits inside

Tail bearing
outer tube
is 42 mm

Tail boom welded to outer bearing tube

The steel plate fits into the slot in the
tube, restricting the movement of the
tail hinge bearing.

Slot in tube

End view showing the angle
subtended by the slot

Fig. 8.3 Fabricating a yaw bearing and tail-hinge from steel pipe.

——Table 8.2 A Rough Guide to Steel Wire Rope——

Diameter	SWL	Proof load	Breaking load
6mm	400kg	800kg	2 tonnes
8mm	700kg	1.4 tonnes	3.5 tonnes

Some rules for guying your tube:

1. Attach the topmost guys as close as you dare to the blade tips. The further your guys are below the windmill, the more the tower is likely to flex. But beware! The blade tips will be deflected backwards. Too often, the blades have hit the guys!

2. Put the guy anchors as far out as possible. A small guy radius means a very tight guy, and this will put a buckling load on the tower.

3. Stiffen the tower with intermediate guys at six metre (6 m) intervals between the base and the top guy level.

The minimum number of guys you can use at each level is three, but four is a much more convenient number if the tower is to be tilted up and down to the ground. Six or eight may be advisable if using cheap materials such as fencing wire.

Guy materials

For a small windmill (one metre diameter) on a temporary site, fibre rope is ideal for guys. Blue polypropylene rope is cheap, strong and easy to use. Make sure you know your knots, and keep cattle away: they love to chew on rope!

Fencing wire is a low cost material, suitable for guying windmills up to about two and a half metres (2.5 m) rotor diameter. Specialist tools and skills are required for handling fence wire, though.

Wire rope is the best material for guys on larger machines or for unskilled persons. Buy it from a specialist supplier who can provide all the fittings, including shackles, and certificates which indicate the safe working load.

Your supplier will also be willing to cut the rope into lengths, with a neatly crimped eye on one end. The normal practice is to leave the other end 'plain' so that you can adjust the length on site. However, it is cheaper to make up your own guys, cutting the rope with a hammer and chisel, and putting your own eyes on both ends.

An Eye on Steel Wire Rope

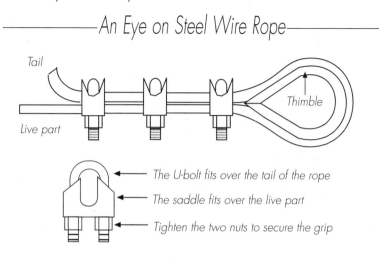

Fig. 8.4 Making an eye at the end of a steel wire rope. Use at least three grips to secure the tail to the live part. In reality they are often more widely spaced.

To make an eye at the end of a flexible steel wire rope, you need a thimble, and at least three rope grips. The thimble is an insert which supports and protects the inside of the rope loop. The grips (also variously known as Bulldog grips, cable clamps or Crosby clips) secure the tail of the rope to the live part (see Fig. 8.4).

As you tighten the grip, the U-bolt will cut into the rope, so it is always fitted to the tail, not the live part. The saddle fits on the live part.

Use turnbuckles or rigging screws to tighten the guys (Fig. 8.5). Avoid the kind with hooks on the ends. The hooks sometimes bend or break under load, or the rope can slip off if it slackens. Use turnbuckles with eyes or jaw ends (or both).

Eye ends are ideal for attaching shackles. You can even fit a thimble direct to an eye, if you open it carefully. The best way to open a thimble is by twisting it (Fig. 8.5).

Jaw ends are very useful, as you can easily insert a thimble, or fit the jaw over a steel plate with holes for bolts (see Fig. 8.6). A single steel plate at the anchor can hold several turnbuckles; one for each guy. Side guys (see later) will need shackles for extra flexibility.

Tensioning Devices

Fig. 8.5   Turnbuckle with hook and eye ends. Avoid the hook ends.

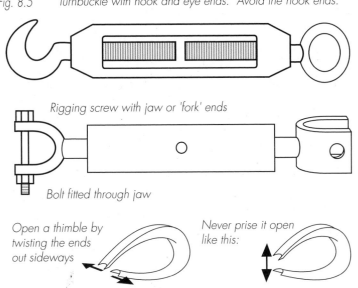

Rigging screw with jaw or 'fork' ends

Bolt fitted through jaw

Open a thimble by twisting the ends out sideways

Never prise it open like this:

Turnbuckles on a Steel Plate

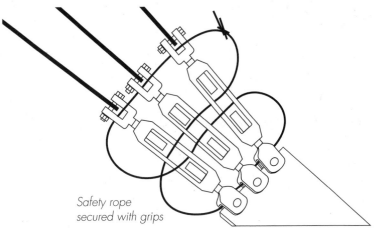

Safety rope secured with grips

Fig. 8.6 Jaw-ended turnbuckles on a steel plate.

Jaw Ended Turnbuckles

The chains are attached to a 'Dead Man'.

Lock the turnbuckles against coming loose accidentally (or through vandalism) by threading a spare piece of wire rope (the 'safety rope') through the body of the turnbuckle(s), and also through the thimbles or jaws at each end. Tie the ends of the safety rope together with rope grips.

A low-cost alternative to using turnbuckles is the 'stay-rod', used everywhere for electricity poles. This is a long, thick steel bar which connects a buried anchor (or 'dead man') to the guy. At the top, the bar is threaded so that you can adjust the position of a large 'bow' which holds the guy rope.

If you cannot afford to buy any special fittings, two M12 galvanised threaded rods make a good, adjustable connection to a buried steel pipe (Fig. 8.7).

The guy rope requires no thimble for this arrangement. You can also avoid the need for a thimble at the top end, by passing the guy rope around the tower, rather than using a shackle. Take care that vibration of the guy does not fray it where it meets the tubing.

Homemade Adjustable Anchoring

Fig. 8.7 *Using two threaded rods for an adjustable deadman attachment*

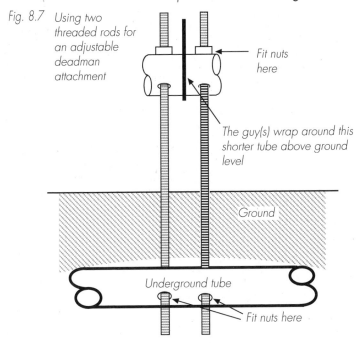

Fit nuts here

The guy(s) wrap around this shorter tube above ground level

Ground

Underground tube

Fit nuts here

A Dead Man with Two Rods

Ready for burial.

Concrete Anchor-Block

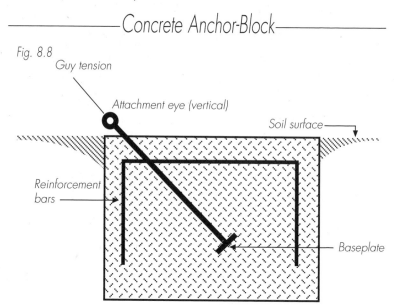

Fig. 8.8

Guy tension

Attachment eye (vertical)

Soil surface

Reinforcement bars

Baseplate

Anchors

Concrete blocks

The neatest anchor is a mass of reinforced concrete set into the ground. A cubic metre of concrete weighs over two tons, and this will normally be adequate for rotors of up to 3 metres diameter (you should do your own calculations, based on the actual guy radius). Concrete anchors are usually cast *in situ* using 'readymix' poured into a hole.

You need reinforcement bars across the top and down the sides. They must not penetrate the surface of the block or they will corrode (see Fig. 8.8).

Eyebolts or plates are embedded in the anchor to provide attachments for the guys. If an anchor is also to be used for winching purposes then it is a good idea to have a separate lug for the winch. The line of the guy should point toward the centre of gravity of the anchor block.

Rock bolts

Where there are large areas of bedrock exposed at the site, then rock bolts may be cheaper than concrete anchors. Probably the easiest type of rock bolt to fit is the kind supplied with epoxy resin in glass vials. You drop a vial into the hole and hammer the bolt in after. If in doubt about the strength, apply a proof load to the bolt of (say) twice the working load. This is very reassuring.

Dead men

A buried object (known as a 'dead man') is an excellent anchor. Wire rope and fencing wire are not suitable for use below or at ground level because they will corrode. You need something thick. The guys are attached to a chain or bolt rising from the anchor to above ground level.

Dig a hole and bury a piece of steel pipe or treated timber. The depth will depend on the forces involved, but between half a metre and a metre will usually be ample. If you are not sure of the holding capability of a ground anchor, you can apply a 'proof load'.

Fence posts or other stakes

For windmills under 2.5 metre diameter with low guy angles (where guy radius exceeds tower height), you can simply use stakes driven into the ground as anchors. Holding will depend entirely on soil conditions.

Use a thick piece of galvanised steel pipe or a treated fencing post, driven well in at right angles to the line of the guy. A vehicle half-shaft also makes a handy anchor-stake (complete with flange and holes for shackles). Steel anchor stakes have the advantage of being easier to drive in to a good depth.

Tilt-up towers

Tilt-up towers are hinged at ground level, for easy erection. Use a gin pole to provide enough radius for the lifting rope (Fig. 8.1). Some towers will need several guys from the gin pole to the tower to prevent the tower from sagging in the middle.

Figures 8.9a and b overleaf show some typical hinged-base fabrications. Do not forget to allow an exit for the cable.

Guying tilt-up towers

You will need four guy anchors, equally spaced at right angles

Hinged Base Fabrication

Fig. 8.9a

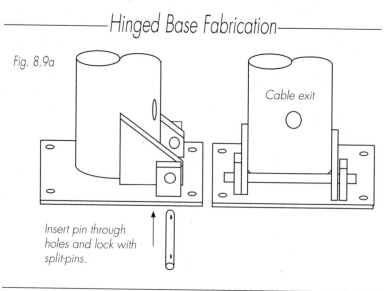

Cable exit

*Insert pin through
holes and lock with
split-pins.*

Drilled and Bolted Base

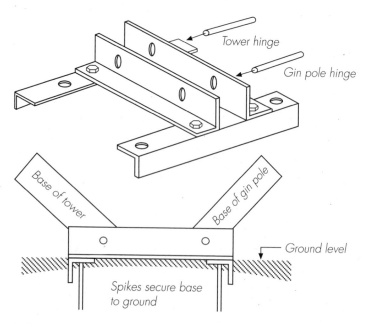

Tower hinge

Gin pole hinge

Base of tower

Base of gin pole

Ground level

*Spikes secure base
to ground*

Fig. 8.9b Hinged base using drilled and bolted steel angle.

Anchor Layout for a Tilt-Up Tower

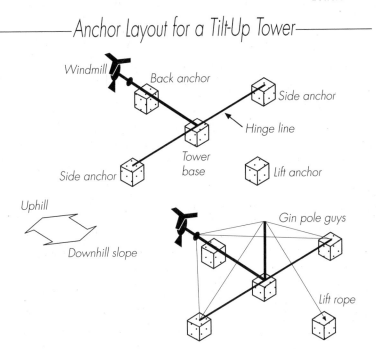

Fig 8.10 The lift anchor is downhill and the side anchors and tower base are on the same contour.

(see Fig. 8.10). The two anchors in line with the hinge on the tower base are the 'side anchors'. (These are also used to guy the gin pole.) The one from which you lift the windmill is the 'lift anchor'. The one which the windmill lies on at the start is the 'back anchor'.

If the ground slopes, then the side anchors should be on the same contour as the tower base. The lift anchor is downhill, and the back anchor is uphill.

Make the hinge line straight

The line between the two side-anchors *must* pass *exactly* through the hinge, *not* through the centre of the tower base itself, if the hinge is offset as is often the case. This is more important than the exact levels of the various anchors. A line of sight from one side-anchor attachment to the other must pass through the hinge pin.

If this line is not straight, then the side guys will tighten either

────────Pole Length Equals Guy Radius────────

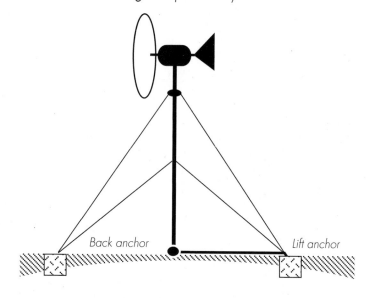

Back anchor   Lift anchor

Fig 8.11 When erect, fasten the gin pole to the lift anchor.

as it is raised or lowered. This is potentially very dangerous. Very large forces are produced, and the people involved are not usually looking at the side guys, but rather at the lifting rope and the windmill. The side guys can become so tight that they break, with disastrous consequences!

Gin pole length

The longer the gin pole, the easier the lift, so make it as long as you conveniently can. If the guy radius is less than the tower height, then make it long enough to reach the lift anchor, so you simply need to connect the gin pole (complete with guys) to the anchor (Fig. 8.11).

Hints for safe erection of tilt-up towers

Safety must always be the first consideration. Take things slowly. If great strength is required then you are doing something wrong.

- Ensure that the side anchors are in line with, and level with,

Using a Gin Pole

A pulley on a gin pole and 12m tilt-up tower.

the tower base hinge.
- Lay the tower out at right angles to this line uphill, and assemble the base hinge.
- Attach the side guys to the side anchors and adjust them.
- Lay the gin pole out on the hinge line and attach the guys from it to the tower before erecting the gin pole, sideways.
- Find the approximate length for the back guys (by laying them out to a side anchor) and attach them to the back anchor.
- Attach any remaining guys to the tower, in readiness for connecting them to the lift anchor after the lift.
- Check everything thoroughly, especially the lifting arrangements. All shackles must be tight and all nuts locked.
- You may wish to attach a back-rope to control the tower (see later).
- Lift the tower without the windmill, as a trial run.
- Lift slowly and deliberately, without jerking. Stop from time to time, and check the guy tensions. No-one must walk beneath the tower during erection.
- When nearly erect, stop and check visually that the back-guys are not kinked or twisted in any way.

The 'Tirfor' type Hand Winch

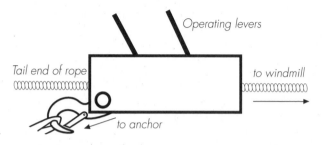

Fig 8.12 *The rope passes right through and out the back.*

- You may need to control the tower with the back-rope to prevent the gin pole from overbalancing it, and jerking the back anchor.
- Adjust all the guys to your complete satisfaction before lowering the tower, and fitting the windmill.

To adjust the guys

All the turnbuckles need to be fully extended at first. If there is more than one set of guys, start by adjusting the bottom ones. Place a spirit level against the tower. When the bottom part of the tower is vertical, you can do the rest by sighting up along the tower, and checking that it is straight.

Some ropes will need tightening, some may need slackening. Adjust them one at a time. Loosen the rope grips, and pull up as much of the slack as you can by hand. When the ropes are as tight as you can make them, and the tower is fairly vertical, tighten the turnbuckles.

Do not over-tension the guy ropes, as this puts an unnecessary extra load on everything. Start by rotating the turnbuckles by hand, followed by a few turns with a spanner. If the tower is not vertical, then you can adjust it with the turnbuckles. You will need to loosen one guy before you tighten the opposite one.

Winches and pulley systems

If you have to buy something, the best investment is probably a hand winch of the 'Tirfor' type. (There are other brands, but they are all similar, and the name is loosely applied to all.)

The Tirfor is hard to beat for erecting tilt-up towers, because it is slow and fail-safe. There are two levers you can use: one for pulling in, and one for paying out the rope. If you stop moving the levers, the rope is locked in position. The winch works 'hand over hand' so there is no limit to the length of rope it can process, given time (see Fig. 8.12).

Tirfor machines are available in a range of sizes, each designed for a particular 'safe working load' but you can increase this pulling capacity, if necessary, by using pulley blocks. This will require a longer rope than a direct pull would, though.

Towing with a vehicle

This is an attractive way to lift things, because most people have a vehicle available at no extra cost. It is not as controlled as using a Tirfor type machine though, and it may end up costing you more, if you are not very careful!

The problem boils down to the vehicle being too fast in most cases, so things go better with:

- long gin poles;
- pulley systems which double the rope;
- low gear-ratio vehicles.

So we come to the end

There is not much to beat the magic moment when your windmill starts to turn and power flows into your battery for the first time. If there is no wind, you can take this as an excellent omen. This is your first taste of the loveable perversity of wind energy. One of the advantages of a successful windpower installation is that the weather instantly becomes flat calm for weeks!

I hope you have enjoyed this book and learned some new things which you can put to use. Much has been left out for lack of space, so watch out for future publications. I am always happy to discuss details on the phone or by e-mail, but unwilling to answer the simpler question "Please tell me how to build a windmill". The answer is not simple, but I hope you agree it is worth finding out.

Glossary

a.c.: electric current which alternates its direction, to and fro, as supplied by the mains grid or by an alternator.

Acid: the electrolyte of lead acid batteries is a mixture of sulphuric acid and water, which needs to be topped up with pure water occasionally. Never handle acid!

Airfoil section: a cross-section of a wing shape, designed for low drag and good lift.

Air gap: the space between the rotor and the stator of a generator.

Alternator: a generator which produces alternating current (most do).

Amps: units of electric current flow.

Anemometer: an instrument for measuring windspeed. It usually has three or four 'cups' spinning around on a vertical shaft.

Angle of attack: the angle of the relative air-flow to the chord of the blade section.

Armature: another word for the rotor of a dynamo.

Axis: the invisible line at the centre of a rotating object's movement.

Balancing: spinning objects will wobble excessively if the centre of gravity is not exactly on the axis (centre) of rotation. Balancing a windmill rotor, so that it will sit level on a spike or thread at the exact centre, will prevent vibration.

Batteries: stand-alone systems use rechargeable batteries, usually of the lead-acid type, similar to, but not the same as, car batteries.

Betz (Albert): the man whose name is associated with the theorem which tells us how much to slow down the wind in order to extract the most power from it.

Blade: the wing, sail or paddle of the rotor that catches the wind.

Brake switch: a switch which short circuits the output from a permanent magnet alternator in order to stall the rotor blades and stop the windmill.

Bridge rectifier: a rectifier circuit employing several diodes to give a continuous d.c. output.

Brushes: carbon blocks which rest on revolving surfaces and pass current between the stator and the rotor.

Centrifugal force: the outward 'force' experienced by spinning objects.

Chord: the width of the blade from leading to trailing edge.

Cogging: a 'lumpy' torque caused by 'teeth' on the stator passing poles on the rotor.

Coil: a length of enamelled copper wire, wound around a former.

Commutator: a segmented copper cylinder used with brushes to create d.c. output from a dynamo.

Coning: hinged movement of the blades of a HAWT, in a downwind direction. Centrifugal force can prevent this movement from going too far.

Copper loss: see 'resistance'.

Cp: abbreviation for power coefficient.

Current: the flow of electricity, measured in amps. Power depends also on voltage.

Cut-in: the lowest windspeed (m/s) or shaft speed (rpm) at which output power is obtained.

d.c.: direct current which always flows from supply positive to negative, as supplied by batteries (or by a rectifier).

Delta: a type of three phase connection where each phase is connected between two supply wires.

Density: mass per unit of volume. Air weighs 1.2kg per cubic metre.

Diameter: the distance from one side of a circle to the other.

Diode: a 'one way valve' for electric current. Cheap, reliable semiconductor device, rated for particular maximum voltage and current. Needs a voltage (about 0.7V) to push current through it.

Direct drive: drive without gearing. Blades may be mounted on the rotor of the generator.

Downwind: an unusual type of HAWT in which the rotor blades are behind the tower when viewed from the windward side.

Drag: moving air exerts a drag force (on an object) in the direction of relative movement.

Dynamo: a generator which produces d.c. using a commutator.

Earth: the planet where we live, and any wiring connected to this planet for safety purposes.

Electromagnetic induction: the miracle whereby the relative motion of flux and wires creates an electric voltage in the wires.

Emf: electromotive force. The voltage 'behind' a supply, before it suffers volt-drop due to current flow.

Energy: the 'stuff' of heat, light and motion, usually measured in watt-hours or kilowatt-hours (kWh). Force multiplied by distance.

Epoxy: very strong, two part resin, useful for glueing or protecting windmill blades.

Equalising: charging the batteries faster than usual in order to ensure that they all reach full charge.

Excitation: the creation of a magnetic field using an electric current.

Fatigue: the type of failure which occurs as a result of repeated cyclical stress rather than of a single excessive force.

Ferrite magnets: cost-effective but heavy permanent magnets.

Field: magnetic flux (near enough for the purposes of this book).

Flux: magnetic field.

Frequency: the rate of alternation of a.c. The UK grid frequency is 50Hz.

Furl: to reduce the exposure to the wind by facing the rotor 'off' to the side or upward.

Fuse: A thin piece of wire which is used to protect against excessive currents in a circuit. If the current is too large, the fuse will melt.

Gale: windspeed above 14 m/s (31 mph).

Gearing: the use of mechanical drive belts, etc., to increase the rotational speed between the wind rotor and the generator rotor.

Generator: a device which converts mechanical power into electrical power.

Govern: to limit the speed of the windmill by altering the attitude or pitch of the rotor.

Guy radius: the distance from the tower base to the guy anchor.

Gyroscopic force: a spinning object (a rotor) is subject to gyroscopic force when swivelled around. The direction of the force is at right angles to both movements. If the speed is high then so are the gyroscopic forces.

HAWT: horizontal axis wind turbine.

Headwind: the wind an object experiences as a result of its motion, rather than the 'real wind'.

Horizontal axis: the usual type of windmill with horizontal shaft and vertical rotor.

Induction motor: the commonest type of motor for a.c. use. Can be used as a stand-alone generator with capacitors.

Inverter: complicated device which converts d.c to a.c to supply mains voltage appliances.

Kinetic energy: energy carried by a moving object and released when it is slowed down. Depends on the square of the speed.

Laminations: flat sheets of steel built up into a core to carry flux as part of the magnetic circuit.

Leading edge: the edge of the blade which would hit your hand if you put it into the spinning rotor.

Lift: moving air exerts a lift force (on an object) at right angles to the direction of relative movement.

Live: an object charged with a dangerous voltage is described as 'live'.

Load: something which absorbs energy (e.g. an electric light). If a windmill is unloaded, it will tend to over-speed, with alarming results.

Loss: power which the windmill captures but fails to deliver.

M/s: metres per second, speed. Multiply by 2.24 to convert to mph.

Magnetic circuit: the pathway for magnetic flux from one pole of a magnet to the other.

Mains: the national centralised electricity supply grid system.

MCB: miniature circuit breaker. Increasingly used in place of the fuse because it is more effective and easier to reset.

Moment: a force acting at a radius to turn something.

Mph: miles per hour speed. Multiply by 0.447 to convert to m/s.

NACA: a series of airfoil sections for which data is available in many reference books.

Permanent magnet: a magnet which need no current to sustain its field.

Phase control: the amount of current delivered to a load can be adjusted by adjusting the timing of the trigger pulse which fires the triac which feeds the load.

Phase: the timing of the alternations of a.c. A single phase supply has only two wires, but a three phase supply has three, and the voltage fluctuations between them are spread out over time.

Power coefficient: the fraction of the power in the wind which is actually delivered, after losses. Often called the efficiency.

Power: energy per second, measured in watts. Force multiplied by speed. Torque multiplied by angular speed. Volts multiplied by amps.

Pulley: a wheel fitted to a shaft to carry a belt for drive purposes.

PWM: pulse width modulation is a way to adjust the current to a load by rapid switching.

Radius: the distance from the centre (to the outside of a circle, to a 'station', or to the line of action of a force).

Rare-earth magnets: expensive but powerful permanent magnets.

Rated windspeed: the windspeed at which the windmill delivers its rated power output.

Rated: the designer's intended value (not always the reality). If rated voltage is 12 volts for example, then equipment will operate best between 11 and 14 volts.

RCD: 'residual current device' which switches off the supply in the event that a small current flows (dangerously) to earth.

Rectifier: a device which converts a.c. to d.c. using diodes.

Regulator: system voltage is regulated by limiting the battery charge current.

Relative wind: the wind 'seen' by the blade, as a result of the actual wind and the headwind due to its own motion.

Resistance: the voltage per amp required to push a current through (for instance) a wire. The voltage drop, times the current, is the copper loss (watts). Resistance of a single wire with cross sectional area 1 mm^2 is approximately $2/100$ volts per amp, per metre length.

Root: the inner end of the blade nearest the hub.

Rotor: something which rotates. Either the moving part of a generator or the blade assembly of a windmill, depending on the context.

Rpm: revolutions per minute. Equals sixty times the blade tip speed divided by rotor circumference (diameter times 3.14).

Runaway: an undesirable phenomenon where the windmill rotor is free of load and so accelerates to its maximum tip speed ratio, causing noise, and possibly damage to the machine.

Savonius: a slow type of VAWT.

Section: see 'airfoil section'.

Servo-motor: a motor which controls rather than drives things.

Setting angle: the angle which the blade chord makes with the rotor plane at any given station; also referred to as 'blade angle' or 'pitch'.

Short circuit: a circuit with virtually no resistance, which draws a dangerously large current from the supply.

Shunt regulator: a shunt is a dump or ballast load which is used to divert current away from the battery and thus regulate the voltage.

Sliprings: smooth copper rings used with brushes to connect the rotor to the stator.

Stall: a loss of lift and torque which occurs when the speed drops too low, and the 'angle of attack' becomes too large.

Stand-alone system: an electricity supply which is independent of the mains grid.

Star: a type of three phase connection where each phase is connected to one supply wire and a common 'neutral' point.

Start-up: the windspeed at which the rotor starts up and spins. Often the start-up is above the cut-in, but once started the windmill will continue to run.

Station: one of several positions along the blade at which the shape is defined in a specification.

Stator: the part of a generator which stays still.

Storm: windspeed above 28 m/s (63 mph). Rare on land.

Tail: the vane at the back of the windmill which holds it facing into the wind.

Thickness: the fattest part of the section, measured at right angles to the chord line.

Thrust: the force which the wind exerts on the rotor as a whole.

Tip speed ratio: the speed of the tip of a rotor blade divided by the windspeed at the time. Rotors will function best at a particular 'design' tip speed ratio.

Tip: the extreme outer end of the blade.

Torque: turning force. Force times radius of action. Like moment.

Tower: the structure which supports a windmill (often a guyed pole).

Trailing edge: the edge of the blade which the wind touches last. This edge is thin to avoid creating turbulent wake eddies.

Transformer: a simple device for converting power from one voltage to another, but it only works on a.c.

Tsr: 'tip speed ratio'.

Twist: blade setting angle changes between the root and the tip.

Upwind: the usual position of the windmill rotor is upwind of the tower.

VAWT: vertical axis wind turbine.

Vertical axis: a type of windmill where the shaft is vertical and the blades revolve in a horizontal direction.

Voltage: the 'push' behind an electricity supply.

Windshear: the rate at which windspeed changes with height above the ground.

Windmill: informal term for any machine which harnesses the power of the wind.

Windpower: power obtained by capturing and converting the kinetic energy of moving air.

Yaw: rotation of the whole windmill on the tower top, to face changing wind directions.

Windpower Equations

(These equations are in a form suitable for use in a computer programme.)

* means 'multiply by', / means 'divide by', and ^ means 'to the power of' the first thing (only) that follows the symbol.

Variable	Symbol	Units	Notes or Equation
Pi	Pi	none	Pi = 3.14 (geometrical constant)
Density of air	rho	kg/m^3	rho = 1.2 (temperature dependent)
Power coeff.	Cp	none	Cp < 0.6, say 0.15
Windspeed	V	m/s	try 10m/s (= 22mph)
Diameter	D	metres	D = (P/(Cp*rho/2*Pi/4*V^3))^0.5
Power	P	watts	P = Cp*rho/2*Pi/4*D^2*V^3
Mean windspeed	Vm	m/s	Vm = around 5 m/s usually
Mean power	Pm	watts	Pm = 0.14*D^2*Vm^3 (approx.)
Tip speed ratio	tsr	none	tsr = rpm x π x D/60/V
Shaft speed	rpm	rpm	rpm = 60 x V x tsr/(π x D)

Blade Design

Radius (station)	Rs	metres	distance from central axis
No. of blades	B	none	An integer (3 is best)
Lift coeff.	Cl	none	Cl = 0.8 say (Alpha dependent)
Angle of attack	Alpha	degrees	Alpha = 4 say (chosen for best lift/drag)
Setting angle	Beta	degrees	Beta = ATAN(D/3/Rs/tsr)*57.3-Alpha

(NOTE:'ATAN( )' is a software function giving 'the angle whose tangent is' in radians.)

Chord width	Cw	m	Cw = 1.4*D^2/Rs*COS(Beta/57.3)^2/tsr^2/B/Cl

Variable	Symbol	Units	Notes or Equation
Alternators			
No. of poles	Np	even no	$Np = 120 * f / rpm$
Frequency	f	Hz	$f = Np * rpm / 120$
Length of airgap	Lgap	m	length parallel to axis
Diam. of airgap	Dgap	m	$Dgap = $ radius from shaft axis $* 2$
Area of air gap	Agap	m^2	$Agap = Lgap * Pi * Dgap$
Cut-in speed	Crpm	rpm	$Crpm = 12$ volt cut-in speed
No. of coils		Ncoils	$Ncoils = $ no. of coils in series
Turns/coil		Nturns	$Nturns = 1200 / Agap / Crpm / Ncoils$

(Note this is only approximate. Increase by 50% for very large airgaps

Copper loss in wires/cables

Copper diam.	Wdiam	mm	$Wdiam = $ diameter of wire
Copper area	Warea	mm^2	$Warea = Wdiam^2 * 0.785$
Twin cable len.	Tcl	m	if investigating cable
Single wire len.	Swl	m	$Swl = Tcl * 2$ (or $Swl = $ length of wire in coils)
Single wire res.	Swr	ohms	$Swr = Swl / Warea / 50$

(NOTE: Resistance is at copper temp. = 50 deg.C, increasing by factor of.004/degC.)

Current	I	amps	$I = $ current through coil or cable
Volt drop	Vdrop	volts	$Vdrop = Swr * I$
Power lost	Ploss	watts	$Ploss = Vdrop * I$

Tail Vane side force

Area of vane	Avane	m^2	$Avane > D^2 / 40$
Side force	Fside	kg	$Fside = Avane * V^2 / 16$
Rotor thrust	Frotor	kg	$Frotor = D^2 * V^2 / 24$

Worked examples of how to use the windpower equations for windmill design.

The equations use variables to calculate answers which themselves may become variables in future equations. The notes suggest values for the first four variables: Pi, rho, Cp and V.

Say we are planning a 300 watt machine. If we have not already decided on a rotor diameter from reading chapter one, then:-
$$D = (P/(Cp*rho/2*Pi/4*V^3))^{0.5}$$
$$= (300/(0.15*1.2/2*3.14/4*10^3))^{0.5} = 2.06 \text{metres}$$

Conversely, if we have already chosen a 2m diameter then the power output in a 10m/s wind will be:-
$$P = Cp*rho/2*Pi/4*D^2*V^3$$
$$= (0.15)*(1.2/2)*(3.14/4)*4*1000 = 283 \text{ Watts}$$
(Input variables are approximate, so round the answers off to 2m and 300 watts.)

The mean power output is used to judge what we can run off the wind
turbine. If we have a good site with 5m/s mean windspeed, then:-
$$Pm = 0.14*D^2*Vm^3 = 0.14*2^2*5^3 = 70 \text{ watts on}$$
average.
(For example, a 240 watt load for 7 hours per 24 hour day, uses $(7/24)*240 = 70$ watts average power. Ignoring battery losses!)

Now suppose we find a generator which can produce 300 watts at 1000 rpm. We can calculate the rotor tip speed ratio:-
$$tsr = rpm*Pi*D/60/V = 1000*3.14*2/60/10 = 10.5$$

Reading chapter three, we decide this is too tricky to build, and instead we calculate the rpm we can get with tsr = 6:-

rpm = 60*V*tsr/(Pi*D) = 60*10*6/(3.14*2) = 573 rpm

All this is based on a 10m/s windspeed. We must also consider the power and speed conditions at low winds like 3m/s:-

P = Cp*rho/2*Pi/4*D^2*3^3

= (0.15)*(1.2/2)*(3.14/4)*4*27 = 7.6 Watts

rpm = 60*V*tsr/(Pi*D) = 60*3*6/(3.14*2) = 172 rpm

(So, the generator must produce some power at speeds under 200 rpm, if it is to work well in low winds.)

The **'Blade Design'** section suggests the shape of the blade at each station (see p. 38). Say we use 5 stations along the blade length, at radius 'Rs' = 0.2, 0.4, 0.6, 0.8, and 1.0 metre. We choose B = 3 blades, 'Cl' = 0.8 and 'Alpha' = 4. At each station we shall find the setting angle 'Beta', and the chord width 'Cw' (pages 39-40).

The function 'ATAN()' works on most computer software to give an angle in radians. Multiply by 57.3 to convert this into degrees. We can also find the answer on a scientific calculator by pressing 'INV' and then 'TAN'. If the answer is in degrees, there is no need to multiply by 57.3.

At the tip-of-the-blade station, Rs = 1, so:-

Blade setting angle Beta = ATAN(D/3/Rs/tsr)-4

= ATAN(2/3/1/6)-4 = ATAN(0.111)-4 = 6-4 = 2 degrees

Chord width Cw =

1.4*D^2/Rs*COS(Beta/57.3)^2/tsr^2/B/Cl

= 1.4*4/1*COS(6)^2/36/3/0.8 = .064 metres

The tip chord is 64mm wide and makes an angle of 2 degrees to the rotor plane. At other stations (Rs<1) the chord and the blade angle are larger. Using Rs = 0.2 in the calculations, we find that the chord width becomes very wide at the root, and the blade angle very coarse. It is sensible to slim this down, and linearise the blade somewhat (p.41). We have to follow the calculated answers more closely in the outer part of the blade where most of the real action takes place.

From Beta and Cw, you can work out all the data for carving the blades, as in table 4.1 on page 51. The 'drop' (table 4.1 third column, also fig 4.2 page 53) can be computed from the above data. Simply multiply the chord width Cw by SIN(Beta). If using a computer which expects radians, divide Betaby 57.3.

> 'drop' = Cw*SIN(Beta/57.3)
> 'Thickness' = 0.15*Cw (or you can say = 0.12*Cw at the tip).

In the **Alternators** equations on page 147, frequency (in Hertz) is related to number of magnet poles, and to shaft speed (in rpm). For example a 10 pole alternator running at 570 rpm will produce 47.5Hz:-

> f = Np*rpm/120 = 10*570/120 = 47.5

Conversely, you could use frequency readings to compute the rpm. Some multimeters can read frequency, and can therefore be used to measure shaft speed, when connected to the AC output of an alternator (rpm = f*120/Np).

The next equation helps us to guess the number of turns per coil required in an alternator (before it is built and tested). Suppose we are designing a radial field alternator (p.87). The rotor and stator are both 0.1m long. The diameter is 0.15m. The area of the airgap is:-

> Agap = Lgap*Pi*Dgap = 0.1*3.14*0.15 = .047 square metres

(If the machine has an axial field, you can compute Agap by adding up the areas of magnet faces on one rotor.)

Suppose we want our alternator to begin to charge a 12 volt battery at 170 rpm. Say the stator has 10 coils in series:-

> Nturns = 1200/Agap/Crpm/Ncoils =
> 1200/.047/170/10=15 turns

Notes:

1. This applies to single phase, or to three phase coils in 'delta'

connection. For coils in 'star', divide the Nturns answer by 1.73.
2. Alternators with large airgaps will need more turns - maybe
50% more.

Copper loss in cables and windings depends on the wire length,
the wire area, and the temperature (which rises with heavy
currents). Here we assume 50 degrees C, midway between normal
ambient and hot.

Winding wires are sold by diameter, but cables are sold by
cross-sectional area of copper. To find area of a 0.5mm diameter
wire:-

Warea = Wdiam^2*0.785 = 0.25*0.785 = 0.2 sq mm.

Now suppose we have a 10 sq mm cable, 50 metres long:-
Single wire length Swl = Tcl*2 = 50*2 = 100m
Resistance Swr = Swl/Warea/50 = 100/10/50 = .2 ohms

(At 10 deg C we could divide by 60 instead, giving 100/10/60
= .17 ohms,
while at 100 deg C we could divide by 42, giving .23 ohms.)

Suppose we are using the cable to supply 280 watts at 12 volts
DC. The
current in the cable will be 280/12 = 23 Amps. We can then
work out the volt drop and power loss as follows:-
Vdrop = Swr*I = .2*23 = 4.6 volts
Ploss = Vdrop*I = 4.6*23 = 106 watts
(This is rather a high loss, so we ought to use a shorter or thicker
cable,
or a higher voltage system.)

Finally some rules of thumb for **Tail Vane** design. Our rotor
diameter
is 2 metres, so the vane area should be greater than:-
Avane>D^2/40 = 4/40 = 0.1 sq m minimum.
Side force (in 10m/s wind)
Fside = Avane*V^2/16 = 0.1*100/16 = 0.6 kg
Rotor thrust = D^2*V^2/24 = 4*100/24 = 17 kg.

Access Details

This limited selection of contacts (mainly for UK readers) should be useful, but there must also be other alternatives (not researched here) which may or may not be better. Phone them, ask for a catalogue, or check the url for more information.

Information – Windmill Plans
Kragten Design +31 413 475 770 (High quality information at a high price.)
Jemmett 020 7681 2797 jemmett@compuserve.com
http://ourworld.compuserve.com/homepages/jemmett/booksplans2.html
Scoraig Wind Electric 01854 633286 hugh.piggott@enterprise.net
http://www.scoraigwind.co.uk
Aerodyn Shorepower 01823 666177
USA enthusiast - "Jones, Martin" <mjones@tqtx.com>

Information Websites
Centre for Alternative Technology - http://www.cat.org.uk
UK windspeed map - http://www.bwea.com/noabl/index.html
Homebrew ideas - http://www.awea.org/faq/brew.html
E-mail chat - http://www.egroups.com/list/awea-wind-home/
USA users' magazine - http://www.homepower.com
USA publications - http://www.picoturbine.com
Canada - http://www.windmill.on.ca
Excellent Danish site - http://www.windpower.dk

Windmill Components – Blades
Aerodyn Shorepower 01823 666177
USA lmwands@aol.com

Windmill Components – Chinese Alternators
"W.Peng" <W.Peng@chem.rug.nl>
http://www.yhemdc.com/wind/PMG.htm
Claus Nybroe <windmission@vip.cybercity.dk>

Windmill Components –Dynamos
Roy Alvis 01543 253562

Windmill Components – Motors
PM motors from washing machines: Philips 'series 90 super'
USA pm servo motors + 1 (402) 474 5167

Materials – Electronics, Fuses, Winding wire, Cables, Diodes,
Resistors, etc
Electromail 01536 204555 (the 'RS catalogue')
Farnell 0113 263 6311 http://www.farnell.com (free delivery)
JPR 01582 410055 http://www.jprelec.co.uk (low prices)
Maplin Electronics 01702 554001 http://www.maplin.co.uk
Bull Electrical - http://www.bullnet.co.uk/shops/test/
QVS - 0800 801733 (low priced cable and other electrical goods)
USA www.digikey.com

Materials – Leading Edge Tape
Pegasus Aviation 01672 861578 info@pegasusaviation.co.uk

Materials – Magnets
Cermag 0114 2446136
Magnet Developments 01793 766001
Unimag 01709 829 783
USA www.allcorp.com, www.magnetsource.com,
www.allelectronics.com

Materials – Resins etc
Glasplies 01704 540626
Blake Marine Paints 01703 636373
Everglades Resin 01934 744051

Materials – Wire Rope and Fittings
AP Lifting Gear 01384 74994

System Components – Batteries
Stuart Kirby 01531 650226 (used batteries)
CMP 01438 359090 (high quality batteries)

System Components – Inverters
Wind & Sun 01568 760671 www.windandsun.co.uk
Tornado 01280 848361

UK Small Wind Turbine Manufacturers
Ampair Lumic 01202 749994 http://www.ampair.com
LVM Ltd. 01462 733336 http://lvm-ltd.com
Marlec Engineering Co. Ltd 01536 201588
http://www.marlec.co.uk
Proven Engineering Products Ltd 01563 543020
http://www.almac.co.uk/proven/

Index